AGDEX $\frac{762}{931}$

Safe Tractor Operation and Daily Care

AMERICAN ASSOCIATION
FOR VOCATIONAL
INSTRUCTIONAL MATERIALS
Engineering Center
Athens, Georgia 30602

The American Association for Vocational Instructional Materials (AAVIM) is a non-profit national institute.

The institute is a cooperative effort of universities, colleges and divisions of vocational and technical education in the United States and Canada to provide for excellence in instructional materials.

Direction is given by a representative from each of the states, provinces and territories. AAVIM also works closely with teacher organizations, government agencies and industry.

DIRECTOR

W. Harold Parady, Executive Director, AAVIM

REVISION

J. Howard Turner, Editor and Coordinator, AAVIM

ART DIRECTOR

George W. Smith, Jr., Art Director, AAVIM

GRAPHICS

Thomas B. Brown, Art Associate, AAVIM

ACKNOWLEDGMENTS

This publication was initially prepared in 1959 by **G. E. Henderson,** former Executive Director, AAVIM and Professor Emeritus, Agricultural Engineering, University of Georgia and **C. E. Turner,** formerly of the University of Georgia.

Revisions in 1967 and 1970 were made by **Julian M. Fore,** Professor Emeritus, North Carolina State University.

See page 117 for other acknowledgments.

FOURTH EDITION
Revised 1981

An Equal Opportunity Affirmative Action Organization

Printed in the United States of America

ISBN 0-89606-056-X

Contents

Foreword

This edition of SAFE TRACTOR OPERATION AND DAILY CARE replaces the publication, TRACTOR OPERATION AND DAILY CARE. The key word in the revised edition is "safe." The word "safe" is emphasized because of the increased concern for safety in the workplace. Groups like the Occupational Safety and Health Administration (OSHA), The American Society of Agricultural Engineers (ASAE) and The National Safety Council (NSC) have all been instrumental in emphasizing this awareness for safety. With their help, AAVIM has pointed out in this publication safe practices and procedures needed for you to become a safe, skilled tractor operator.

The material in this publication, as always, is presented in the accepted AAVIM format and is based on the most current information available. Acknowledgment is given to the many individuals, industrial firms and manufacturers for their expertise given in the development of this publication.

Preface

This manual is designed to help prepare a student for the professional entry level as a **tractor operator**. Upon completion of this course of study, a student will be able to **service** and **operate** a farm or industrial (wheel) tractor safely.

The knowledge and skills to be acquired are as follows:

- Express the importance of safe and efficient operation.
- Describe the characteristics of tractors.
- Service tractors.
- Start and stop tractor engines.
- Prepare tractors for operation.
- Drive tractors.
- Hitch and unhitch tractor-operated equipment.
- Operate tractors under load.

This program is designed to meet the present requirements of OSHA and to certify 14 and 15-year-old tractor operators.

Introduction

FIGURE 1. The skilled tractor operator must be competent in all areas of tractor operation.

Almost anyone can "drive" a tractor. However, there is a difference between being able to "drive a tractor" and being a professional "tractor operator."

The **skilled tractor operator** is responsible for more than "just" driving the tractor (Figure 1). It is important to know how to do such daily jobs as checking the oil level, checking the coolant level in the radiator and checking the air cleaner for dirt deposits. It is also important to know what will happen if such jobs are neglected.

The skilled operator knows how to hitch and unhitch equipment, and understands what speeds to use in the field to get the best fuel economy with the least engine wear.

Safety is a primary concern of the skilled tractor operator (Figure 2). It is important to be aware of the risks involved and to be on the alert to avoid accidents.

You can become a skilled tractor operator by completing this course of study. The information is given under the following headings:

I. Understanding Tractor Operation.
II. Preparing Tractors for Operation.
III. Making Operational Adjustments on Tractors.
IV. Driving Tractors without Equipment.
V. Hitching and Unhitching Equipment.
VI. Operating and Transporting Tractors.

FIGURE 2. The skilled tractor operator is concerned with safety and efficiency.

I. Understanding Tractor Operation

Understanding tractor operation is discussed under the following headings:

A. Understanding the Importance of Safe and Efficient Tractor Operation.
B. Understanding Characteristics of Tractors.

A. Understanding the Importance of Safe and Efficient Tractor Operation

"Downtime" is a threatening word to a farmer, grower, rancher, or contractor. It hits where it hurts—in the pocketbook (Figure 3). Some downtime is unavoidable. However, a good tractor operator is a major factor in keeping the tractor rolling.

FIGURE 3. Downtime is expensive whether caused by mechanical failures or by accidents.

The two magic words are **safety** and **efficiency**. As a general rule, the safe operator is an efficient operator, and an efficient operator is generally a safe operator.

From your study of this section, you will be able to **list reasons for practicing safe tractor operation**. You will be able to **explain laws which govern safe tractor operation**. In addition, you will be able to **tell why tractor operation efficiency is important**.

These objectives are discussed under the following headings:

1. Importance of Safe Tractor Operating Procedures.
2. Meeting Requirements of the Law.
3. Importance of Efficiency in Tractor Operation.

1. IMPORTANCE OF SAFE TRACTOR OPERATING PROCEDURES

The importance of safe operating procedures is discussed under the following headings:

a. Operating a Tractor Can be Fun.
b. Who is Concerned About Safety?
c. What are the Hazards?
d. How You Can Avoid Accidents.

FIGURE 4. Operating a tractor can be enjoyable.

FIGURE 5. Almost 1,000 persons lose their lives each year because of tractor accidents.

a. Operating a Tractor Can Be Enjoyable

Operating a tractor can be an enjoyable and profitable experience (Figure 4). In recent years, most of the personal discomforts have been eliminated. For example:

- **Comfortable seats** have been designed.
- **Noise levels** have been reduced.
- **Vibration** is now below maximum safety standards.
- **Umbrellas** and **air-conditioned cabs** are available to help protect you from heat, dust and noise.
- **Safety features** and **warning decals** are installed to help protect you from injury.
- In addition, your tractor may be equipped with a **radio** or **stereo equipment. CB radios** are recommended for safety and convenience.

Why all this emphasis on comfort and safety? The answer is to prevent you from becoming a statistic. Almost 1,000 persons lose their lives each year in tractor accidents (Figure 5).

In addition, 100,000 persons are disabled each year in tractor accidents (Figure 6). You probably know of persons who have been involved in tractor accidents. You may have some scars yourself.

Although every safety precaution possible may be taken, the tractor, by nature, is not a safety vehicle.

FIGURE 6. Unguarded power shafts are responsible for some of the injuries relating to tractor accidents.

FIGURE 7. Manufacturers install caution and warning signs near areas where accidents might occur.

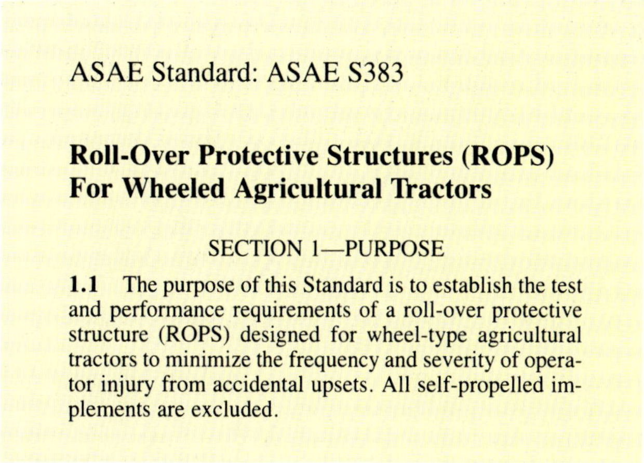
ASAE Standard: ASAE S383

Roll-Over Protective Structures (ROPS)
For Wheeled Agricultural Tractors

SECTION 1—PURPOSE

1.1 The purpose of this Standard is to establish the test and performance requirements of a roll-over protective structure (ROPS) designed for wheel-type agricultural tractors to minimize the frequency and severity of operator injury from accidental upsets. All self-propelled implements are excluded.

FIGURE 8. The American Society of Agricultural Engineers is active in tractor safety.

b. Who Is Concerned About Safety?

Who is concerned about safety? There are many persons and organizations concerned with your safety. Some of them are listed as follows:

- **Tractor manufacturers** have a concern about the safety of persons using their tractors (Figure 7). A person who has been hurt in a tractor mishap is not a very good prospect for future business.
- The **National Safety Council** (NSC) provides safety education programs. It also compiles records of accident statistics (Table I).
- The **American Society of Agricultural Engineers** (ASAE) is active in tractor safety and publishes standards for the design of tractors and implements (Figure 8).
- The **Occupational Safety and Health Administration** (OSHA) of the U. S. Department of Labor issues guidelines under the Occupational Safety and Health Act to assure people of safe working conditions. Tractor safety is included in these guidelines (Figure 9).

TABLE I. CAUSES OF FATAL TRACTOR ACCIDENTS (*National Safety Council*)

Accident	% of Total
Tractor upset	58
Fall from tractor	13
Crushed (other than run over)	9
Run over	8
Motor vehicle collision	6
Power take-off (PTO)	3
Other	3

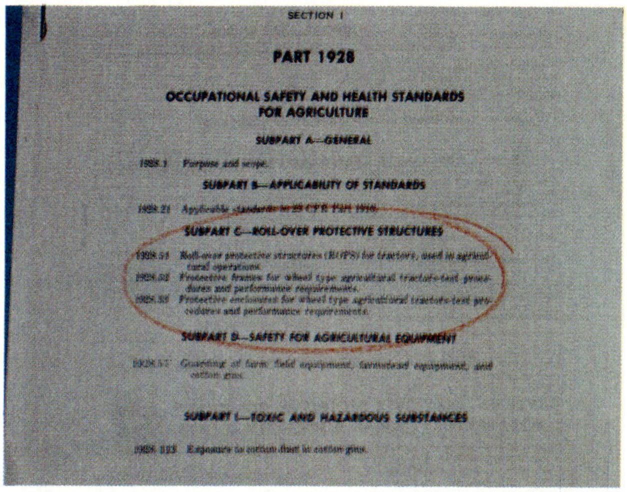
SECTION I

PART 1928

OCCUPATIONAL SAFETY AND HEALTH STANDARDS FOR AGRICULTURE

SUBPART A—GENERAL

SUBPART B—APPLICABILITY OF STANDARDS

SUBPART C—ROLL-OVER PROTECTIVE STRUCTURES

SUBPART D—SAFETY FOR AGRICULTURAL EQUIPMENT

SUBPART I—TOXIC AND HAZARDOUS SUBSTANCES

FIGURE 9. Safety guidelines for tractors are issued by OSHA.

c. What Are the Hazards?

How do people manage to get hurt or killed with tractors? Table I lists the major causes of fatal injuries.

Most tractor fatalities are caused by a tractor overturning. This and other causes of fatalities are listed as follows:

- **Tractor overturns (Figure 10).**
- **Falling from tractor (Figure 11).**
- **Crushing (Figure 12).**
- **Being run over by the tractor (Figure 13).**

  Most of these accidents are a result of first falling from the tractor.
- **Highway accidents (Figure 14).**
- **Getting caught in the power take-off (Figure 15).**
- **Passengers (Figure 16).**

FIGURE 10. Overturns cause over one-half of all tractor-related fatalities.

FIGURE 11. Falling from the tractor is a common cause of injury.

FIGURE 12. Some people get hurt by allowing a tire to fall and crush them.

FIGURE 13. Most people who are run over by a tractor first fall from it.

FIGURE 14. Highway accidents involving tractors are costly.

FIGURE 15. The power take-off (PTO) is the most dangerous moving part of your tractor.

FIGURE 16. Tractors are not designed to carry passengers safely.

d. How You Can Avoid Accidents

What can you do to avoid such tragedies? Just about everything that can be done to make tractors safe has been done. Roll-over protective structures (ROPS) and safety belts are installed on new later model tractors of 15 kW (20 hp) or more (Figure 17). Shields are provided for PTO's and other rotating parts (Figure 18). Decals are placed in strategic places to give operating instructions and warnings against danger (Figure 19). Operating and servicing procedures are given in the operator's manual. Safety procedures are listed (Figure 20). Although manufacturers and organizations provide safety devices and safety procedures, the final responsibilities are yours, the operator (Figure 21).

How does one become a safe and efficient tractor operator? First, recognize the benefits and rewards, such as good health and extra income. Then, make an honest effort to approach every job in the following manner:

- **Be concerned for the safety** of yourself and others.
- **Be alert to danger** at all times.
- **Observe conditions** which threaten safety.
- **Decide on the best alternative** for correcting these conditions.
- **Take a positive action** to remedy the conditions.

Being aware of existing hazards, the concerned person prepares for protection and takes necessary precautions for the prevention of accidents (Figure 22).

FIGURE 17. Tractors of 15 KW (20 hp) or more are now provided with roll-over protective structures (ROPS) and safety belts.

FIGURE 20. Safety procedures are given in operator's manuals.

FIGURE 21. Safety is the responsibility of the tractor operator.

FIGURE 18. Shields protect you from PTO's and other rotating parts.

FIGURE 19. Decals are placed in strategic locations to warn you of danger.

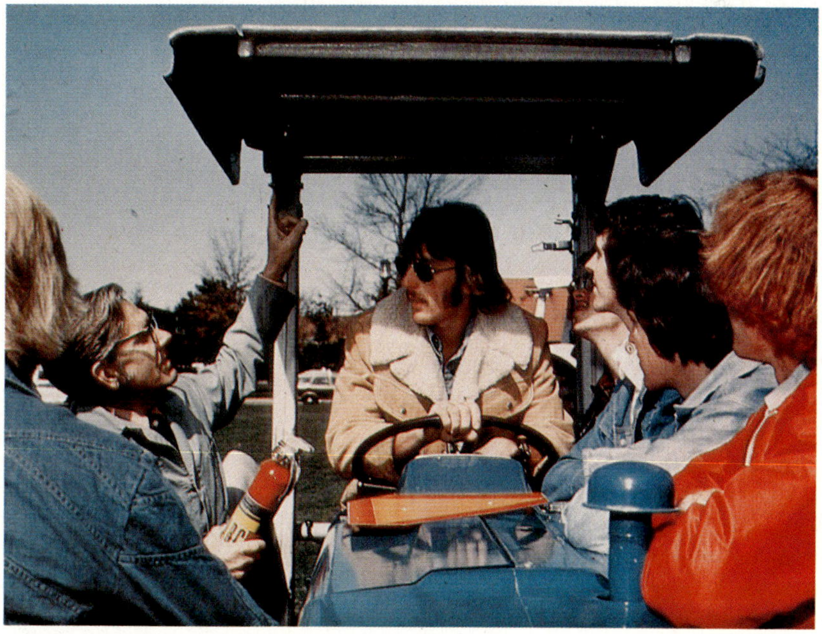

FIGURE 22. Safety is everyone's business.

FIGURE 23. If you are under 16 years of age, you cannot be employed as a tractor operator without special training.

CERTIFICATE OF TRAINING

This is to certify that: Certificate No. USDL ________

________ Date ________
(Name) (Date of Birth)

____
(Address) (Zip Code)

is 14 years of age or more and has successfully completed the ________ (4-H or Vo-Ag)
training program and examinations in:
(Cross out one) 1. Tractor Operation
2. Tractor and Machinery Operation
as specified by the U.S. Department of Labor in the Agricultural Hazardous-Occupations Order (Subpart E-1 of 29 CFR, Part 570) pertaining to the employment of youth under 16 years of age.

Certifying Authority (extension Agent or Vocational Agriculture teacher only) — Person who conducted the training program

FIGURE 24. Sample certificate issued to 14- or 15-year-old youths who qualified for employment as tractor and/or machinery operators.

2. MEETING REQUIREMENTS OF THE LAW

There are two specific regulations governing safe tractor operation:

a. Training Required.
b. The Law.

a. Training Required

In 1967, the Secretary of Labor issued an order that prohibits youths who are below the age of 16 from being **employed** in hazardous occupations in agriculture without special permission. Operating tractors with 15 kW (20 hp) or more and machinery were among those occupations listed. (Youths working on their own family farms are excepted.)

Permission was granted in 1970 to allow 14- and 15-year-olds to be employed as tractor and machinery operators, provided they met certain training requirements (Figure 23). Both Vocational Agriculture and the Extension Service were approved for administering the training programs.

Training for both is described under Subpart E-1, Part 1500, Title 29 of the Code of Federal Regulations, Paragraph 1500.72(b) (3). A minimum of 15 hours of instruction and 9 hours of laboratory work are required for the tractor program. For the machinery course, 10 hours are required, 2 hours of which are laboratory work. These training requirements are outlined in Special Paper No. 8, U. S. Department of Education (formerly an office of the U. S. Department of Health, Education and Welfare).

The student must pass a written examination and demonstrate the ability to operate a tractor and machinery safely. Then, the student is issued a certificate by the instructor (Figure 24).

Persons employing these youths are responsible for the following activities:

- **Keeping a copy** of the certificate.
- **Instructing the youth** on safe and proper operations of the tractor and/or machinery (Figure 25).
- **Checking on the welfare** of the youth at least three times per day—morning, noon and mid-afternoon—and being certain that the youth is following required safety procedures.

FIGURE 25. Persons employing 14- and 15-year-olds in hazardous occupations are responsible for their safety.

FIGURE 26. OSHA enforces safety standards.

b. The Law

Since 1970, OSHA has been the regulatory agency that enforces the safe operation of tractors and machinery (Figure 26). There are certain safety standards established by the Occupational Safety and Health Act that must be met by agricultural employers. The purpose of these standards is to provide workers with a guaranteed maximum of safety and health protection.

OSHA is a federal agency. If a state has standards that are equal to or more stringent than the OSHA standards, the state standard applies. Personnel are trained and sent out by OSHA to make periodic inspections and to determine how well an employed youth is doing. If the agricultural employer is in violation of some or any safety standards, the employer may be fined (Figure 27).

FIGURE 27. The agricultural employer may be fined for safety violations.

Actually, OSHA does not "pick on" agricultural employers but rather regulates their safety standards just as it does for other occupations. However, the accident rate for agricultural workers is generally higher than that of other occupations (Figure 28).

Some of the responsibilities of the agricultural employer to OSHA are to comply with the following:

(1) General Provisions of the Law.
(2) Safety Standards.
(3) General Duty Clause.

(1) GENERAL PROVISIONS OF THE LAW. For the employer to be able to comply with the general provisions of the law, proceed as follows:

- **Become familiar with the standards.** New standards are being developed periodically. Contact your vocational agriculture teacher or county agent and write to your nearest OSHA Regional Office for information.
- **Inform your employees about OSHA.**
- **Check conditions on your farm or ranch for compliance.**
- **Remove or guard against hazards.**
- **Make sure your employees have and use safe tools and equipment and protective gear.**
- **Instruct employees on safe operating procedures.**
- **Report fatal accidents to OSHA within 48 hours.** Also, report accidents that hospitalize five or more persons.
- **Post the OSHA "rights and responsibilities" poster in a prominent place.**
- **Do not discriminate against employees who exercise their rights under OSHA.**

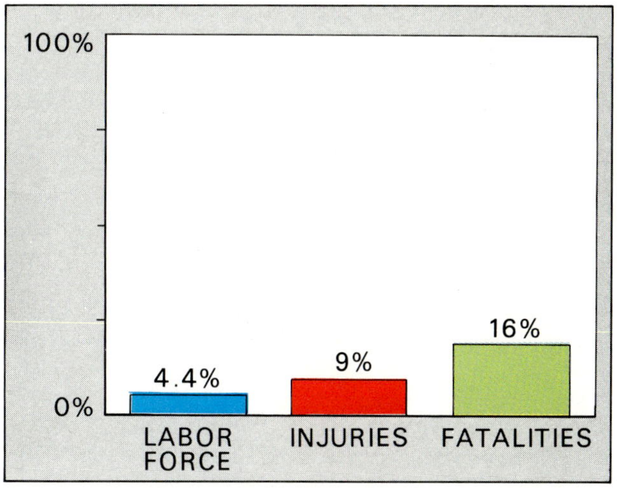

FIGURE 28. Agricultural workers make up sixteen percent of job-connected fatalities and nine percent of job-connected injuries and illnesses. However, they represent only 4.4 percent of the labor force.

- **Post OSHA citations for violations at the place where the violation occurred until the hazard is removed.**

(2) SAFETY STANDARDS. Presently, there are six areas of standards for agriculture covering the following:

- **Temporary Labor Camps.**
- **Storage and Handling of Anhydrous Ammonia.**
- **Pulpwood Logging.**
- **Slow-Moving Vehicle Emblems.**
- **ROPS on Farm Tractors.**
- **Farm Machinery Guarding and Cotton Gins.**

(3) GENERAL DUTY CLAUSE. This clause places on the employer a responsibility for providing safe working conditions in areas which are not covered by standards. If dangerous working conditions exist, you, as an employee, may inform OSHA without fear of retaliation.

Does the employee have any rights and responsibilities? Yes, by all means. No safety program will work without the concern of the employee (Figure 29). If you are the employee, you must obey the safety and health rules and standards. If not, the employer can take disciplinary action against you.

FIGURE 30. Official slow-moving vehicle (SMV) emblem.

The following OSHA standards affecting tractor operation are in force:

- **Slow-moving vehicle (SMV) emblem (Figure 30).**
  The SMV emblem is required on all vehicles which, because of design, move 40 kmph (25 mph or less) on public roads (Figure 31).

FIGURE 29. No safety program will work without the concern of the employee.

FIGURE 31. SMV emblems are required on tractors being driven on public roads.

FIGURE 32. ROPS are required on new tractors of over 15 kW (20 hp), if operated by a hired employee.

- **ROPS (Figure 32).**
  ROPS are required on all tractors of 15 kW (20 hp) or more and manufactured after October 25, 1976, if operated by a hired employee. Exceptions are "low-profile tractors" (tractors with low center of gravity). (Some states may require adding ROPS to older tractors.)
- **Machinery guarding.**
  Twenty percent of all farm-related injuries involve machines. Guards and shields must be in the proper places and decals or warning labels must be placed where needed (Figure 33). Electrical lockouts must be provided where applicable. Employees must be instructed in the proper use of each machine.

3. IMPORTANCE OF EFFICIENCY IN TRACTOR OPERATION

Proper tractor maintenance and operation should always be stressed to increase efficiency, which leads to greater productivity and profit (Figure 34). Insured efficiency means more profit because of the following:

- **Reduces energy consumption.**
- **Saves fuel costs.**
- **Requires less labor.**
- **Reduces downtime.**
- **Results in less waste.**
- **Saves on repair bills.**

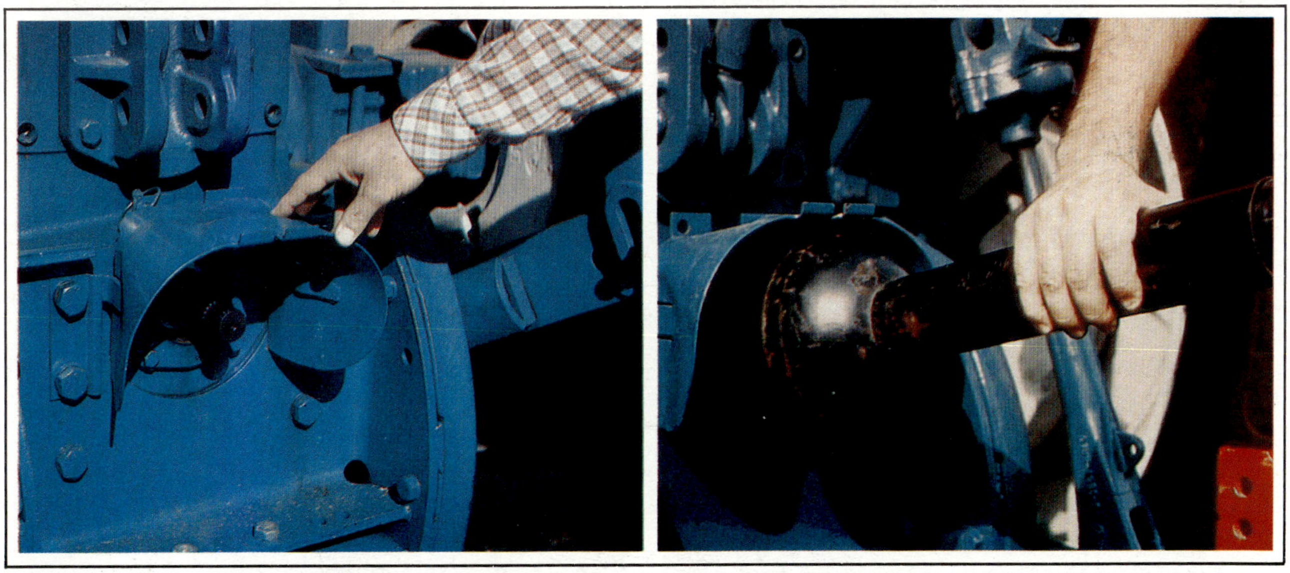

FIGURE 33. All guards and shields should be in place when operating equipment.

FIGURE 34. Proper care and operation increase efficiency and help reduce accidents.

- Helps tractor and equipment to last longer.
- Results in better quality crop.
- Increases yields.
- Results in fewer accidents.

B. Understanding Characteristics of Tractors

Your skill in operating a tractor begins with your knowledge of the tractor and equipment. No tractor should be used for a job that it was not designed to do.

From your study of this section, you will be able to **describe what to look for in the operator's manual**. For example, you will be able **to name and locate instruments and controls**, **to locate vital parts and accessories** and **to note special safety requirements.**

They are described under the following headings:

1. Importance of the Operator's Manual.
2. Becoming Familiar with Instruments and Controls.
3. Observing Vital Parts and Accessories.
4. Noting Special Safety Requirements.

1. IMPORTANCE OF THE OPERATOR'S MANUAL

An operator's manual is developed by the manufacturer for each tractor model (Figure 35). It describes in detail the specifications and operating characteristics of that particular tractor. For example, a typical manual:

- **States the power rating of tractor.**
- **Provides photographs of controls and other important parts.**

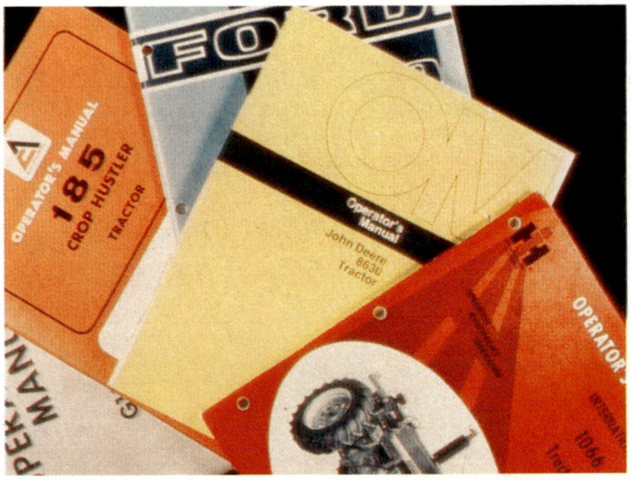

FIGURE 35. Use the operator's manual for your tractor.

- **Explains starting and operating procedures.**
- **Explains gear selection.**
- **Gives service and lubrication specifications.**
- **Recommends tire pressure and wheel weights.**
- **Describes seat adjustment.**
- **Explains wheel tread width adjustment.**
- **Tells how to operate the hydraulic system.**
- **Tells how to connect the remote hydraulic cylinder.**
- **Explains how to hitch and unhitch equipment.**
- **Gives safety precautions.**

2. BECOMING FAMILIAR WITH INSTRUMENTS AND CONTROLS

Before you attempt to service and operate a tractor, become acquainted with it (Figure 36). Different tractor makes and models have instruments and controls located in different places. Starting and stopping procedures may vary. Therefore, it is important to use your operator's manual.

Instruments and controls to look for are as follows:

- **Brakes.**
- **Fuel gage.**
- **Gear selector lever.**
- **Starter.**
- **Oil pressure gage.**
- **Temperature gage.**
- **Ammeter.**
- **Hour meter.**
- **Hydraulic lift controls.**

FIGURE 36. Become familiar with instruments and controls.

3. OBSERVING VITAL PARTS AND ACCESSORIES

Specifications for and proper use of the following are vital to the safe operation of the tractor.

- **Tire inflation.**
- **Wheel weights.**
- **Type of fuel and oil.**
- **Seat adjustment.**
- **Hitch.**
- **Remote hydraulic hose attachment.**
- **Radiator cap (for checking coolant).**
- **Air cleaner.**
- **Crankcase oil dipstick.**
- **Hydraulic oil dipstick.**

4. NOTING SPECIAL SAFETY REQUIREMENTS

Safety begins as you become familiar with the tractor. Look for danger signs such as oil leaks, broken or damaged parts and dirt. Look for special safety features such as the following:

- **Rear view mirror.**
- **SMV emblem for highway driving.**
- **Warning decals.**
- **Clean steps.**
- **ROPS and seat belts.**
- **Securely fastened tools and equipment.**
- **Instructions.**
- **First aid kit.**
- **Fire extinguisher.**
- **Shields in place.**
- **Clear visibility.**
- **Lights for night operation.**

NOTES

II. Preparing Tractors for Operation

Owning and operating equipment is one of the major costs in agriculture. Checking and servicing the tractor before operation is important in keeping these costs down and, more important, in making it safer to operate. A little time spent in checking the tractor before each use will lead to a longer operating life. This regular inspection of the tractor will help to locate small problems needing attention. These problems can be taken care of, thus preventing a major breakdown or serious accident later. The owner or operator of the equipment should keep a log which records maintenance and repair activities performed on the equipment.

You should read the operator's manual to determine the procedures for checking and servicing the tractor before operation. Several of these procedures will need to be performed daily. Other checks are suggested which can be performed at longer intervals. Most manuals generally agree on the items that should be checked at the start of each day. If an operator's manual is not available, the steps given in this section will serve as a guide.

Upon completion of this section, you will be able to **check and service the tractor before operation.** These items are discussed under the following headings:

A. Providing for Safety.
B. Servicing Tractors.

A. Providing for Safety

You must prepare both yourself and the tractor for safe tractor operation. Upon completion of this section you will be able to provide the necessary safety precautions. These are discussed as follows:

1. Personal Safety Precautions.
2. Preparing the Tractor for Safe Operation.

1. PERSONAL SAFETY PRECAUTIONS

Safety skills are a responsibility of the operator. The questions you should ask yourself are as follows:

1. *Are you mentally alert (Figure 37)*?
   If you are overly tired, worried or angry, your chances of having an accident are greater.
2. *Are you dressed properly (Figure 38)*?
   Operating a tractor may not be considered a social event, but there are certain dress requirements in the interest of safety. Wear comfortable

FIGURE 37. You must be mentally alert.

FIGURE 38. You must be dressed properly.

clothing but not baggy clothing that will get caught in moving parts.

3. *Are you wearing safety equipment when needed (Figure 39)?*
4. *Are you looking for possible hazards (Figure 40)?*

FIGURE 39. Wear safety equipment when needed.

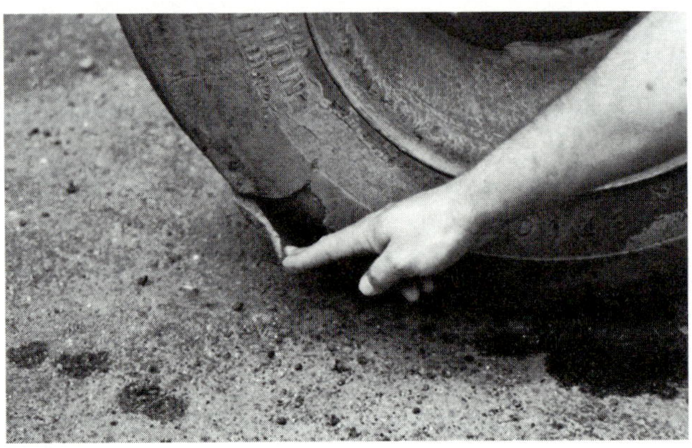

FIGURE 40. Look for possible hazards.

2. PREPARING THE TRACTOR FOR SAFE OPERATION

In addition to the items already listed, check the following:

1. *Read the operator's manual and check the service record (Figure 41).*
2. *Provide for ventilation (Figure 42).*
   Carburetor engines give off carbon monoxide which can cause sickness or death.
3. *Check fire extinguishers for location and condition (Figure 43).*
4. *Read decals on tractor (Figure 44).*
5. *Check first aid kit for location and condition (Figure 45).*
6. *Make sure PTO shield(s) is in place (Figure 46).*
7. *Clear the area of persons, animals and obstacles (Figure 47).*
8. *Clean the steps if equipped (Figure 48).*
9. *Check SMV emblem and warning lights for highway driving (Figure 49).*
10. *Consider other people. Do not allow others on or near the tractor (Figure 50).*
11. *Check lights for night driving (Figure 51).*
12. *Remove or tie down loose tools and equipment (Figure 52).*
13. *Become familiar with controls (Figure 53).*
    Be sure you know how to stop the tractor and engine.
14. *Fasten seat belts before starting engine if your tractor is equipped with seat belts (Figure 54).*
15. *Check visibility (Figure 55).*
16. *Check brakes (Figure 56).*

FIGURE 41. Check the service record.

FIGURE 42. Provide for proper ventilation if in a closed building.

FIGURE 43. Check location and condition of fire extinguishers.

FIGURE 44. Read decals.

FIGURE 45. Locate and check first aid kit.

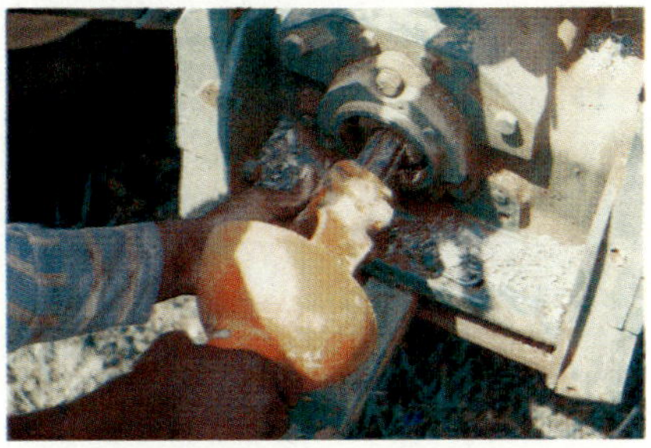

FIGURE 46. Make certain that PTO guards are in place.

FIGURE 47. Watch out for others.

FIGURE 48. Use steps provided for mounting.

FIGURE 49. Check SMV emblem and warning lights.

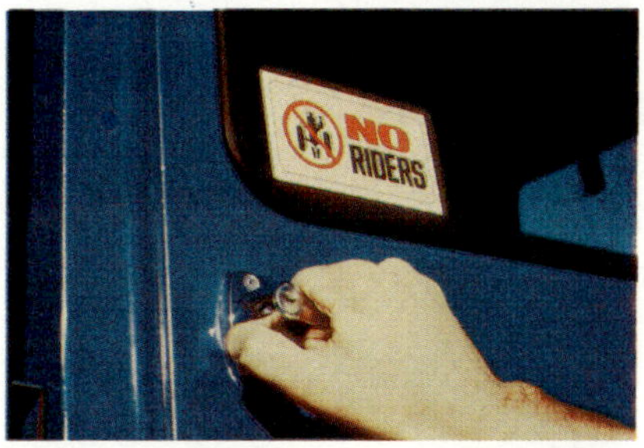

FIGURE 50. Do not allow anyone else on tractor.

FIGURE 51. Check lights.

FIGURE 52. Check for loose tools and equipment.

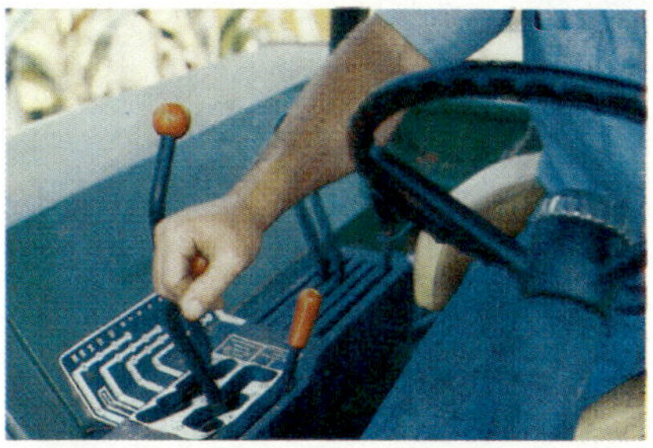

FIGURE 53. Become familiar with controls.

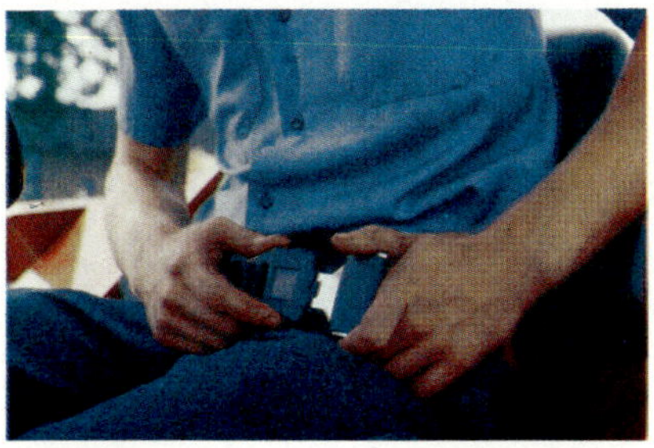

FIGURE 54. Fasten seat belts.

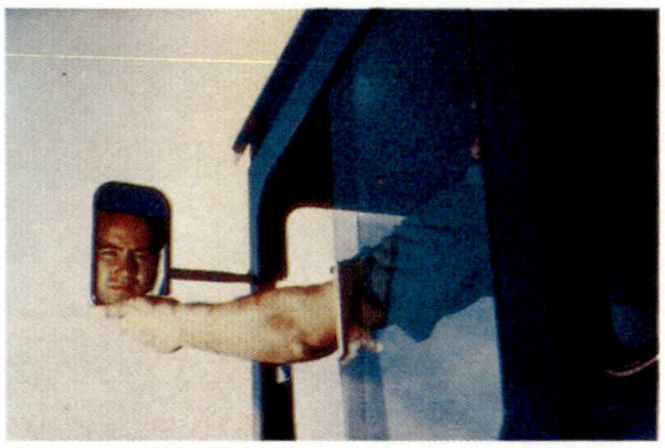

FIGURE 55. Check visibility.

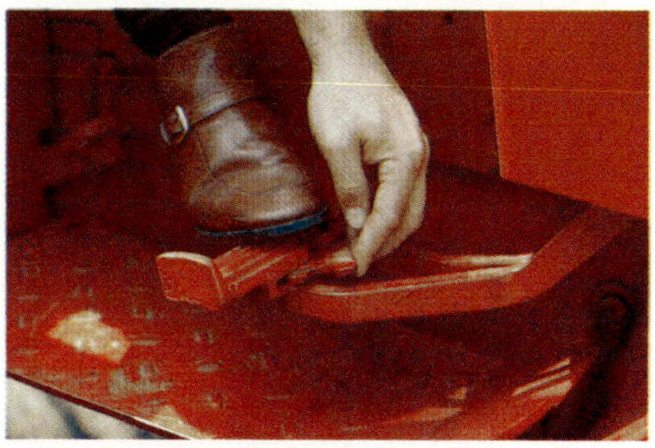

FIGURE 56. Check brakes.

B. Servicing Tractors

Upon completion of this section you will be able to service the tractor properly.

Procedures are given under the following headings:

1. Checking the Crankcase Oil Level.
2. Servicing the Air Cleaner.
3. Servicing the Cooling System.
4. Servicing the Battery.
5. Lubricating the Chassis.
6. Servicing the Diesel Fuel Filter System.
7. Refueling the Tractor.

1. CHECKING THE CRANKCASE OIL LEVEL

You may think of oil as just a lubricant—a means of reducing friction and preventing wear. But it does other important jobs too: it keeps working parts cool, it cushions the pounding action of piston pins and bearings, and it forms a seal between the piston rings and cylinder walls.

If you allow the oil level to get low, it becomes less effective for all of these jobs. A low oil level can be particularly serious if the amount of oil becomes so small that it overheats. Overheating causes oil to oxidize, which means it gets thicker and flows less rapidly. Oxidized oil may also cause varnish deposits, stuck rings and stuck valves. If oxidation continues, corrosive substances that cause rapid engine wear will develop. After the crankcase temperature reaches 82°C (180°F), each additional rise of 12°C (17°F) in temperatures approximately doubles the rate of oxidation.

To check and service the **crankcase oil level**, proceed as follows:

1. *Place the tractor on level surface.*
2. *Apply brakes and stop engine.*
   Allow a few minutes for the oil to drain back into the crankcase.
3. *Check oil level (Figure 57).*
   Remove dipstick. Wipe clean, then recheck.
4. *Note top marking labeled "full" and lower marking lettered "add," "low" or "danger."*
   Some manufacturers make the dipstick short enough so that no oil shows when it is time to add.
5. *Add oil if needed.*
   It is extremely important that you use a **clean container** and **clean oil**. Fill to "full" mark with the type and grade of oil recommended in your operator's manual.

FIGURE 57. Checking oil level with a dipstick.

Do not overfill. Too much oil can cause problems and result in oil foaming.

2. SERVICING THE AIR CLEANER

Checking and servicing the air cleaner on your tractor are two of the most important jobs for you to do. In fact, an Illinois report states that "more tractor troubles result from lack of air cleaner service than any other single cause."

Clean air is important. The common carburetor tractor uses large quantities of air—about 36,000 liters (9,000 gallons) of air for 4 liters (one gallon) of gasoline (Figure 58). Even a small amount of dirt in the air adds up to a considerable quantity during a day's operation. A diesel engine uses even larger quantities of air—especially supercharged engines.

If the dirt is not filtered out, it passes through the intake manifold into the engine cylinders. Here it mixes with oil on the cylinder walls to form a grinding compound. The grinding action takes place on the cylinder walls, piston rings and valves, and in the ring grooves and bearings.

Tractor manufactures equip their tractors with the most efficient air-cleaner units available in an effort to help prevent this unnecessary wear. These air cleaners are about 99 percent efficient. Even with that efficiency, about 34 kg (three-quarters of a pound) of dirt gets into the engine in a season. However, if the air cleaner were only 50 percent effficient, about 16.5 kg (37 lbs) of dirt would reach the engine.

FIGURE 58. With each gallon of gasoline, a tractor uses about 9,000 gallons of air.

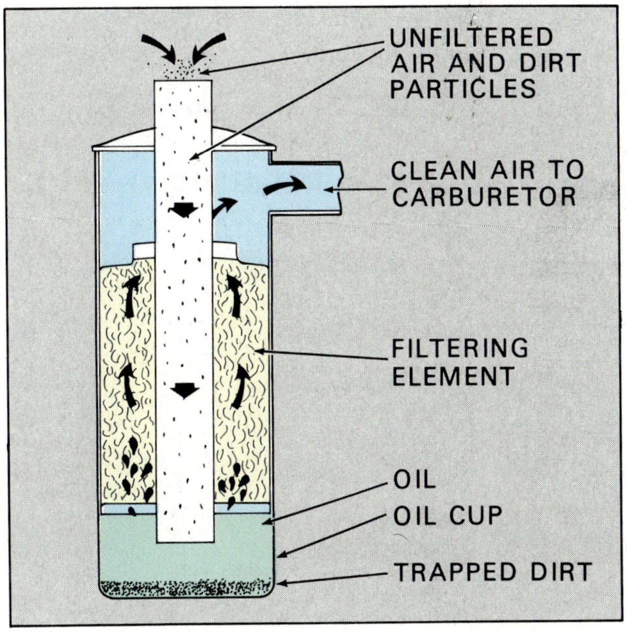

FIGURE 60. Principle of an oil-bath air cleaner.

Servicing air cleaners is discussed as follows:

a. Types of Air Cleaners and How They Work.
b. How to Service Air Cleaners.

a. Types of Air Cleaners and How They Work

There are two types of air cleaners used on farm and industrial tractors:

(1) Oil-Bath Air Cleaner.
(2) Dry Paper Filter Air Cleaner.

It is important that you know the type of air cleaner on your tractor because the frequency and method of servicing each type are different.

With either type, when the engine is running, outside air is drawn through a **screened inlet** or through a **pre-cleaner** (Figure 59) where large particles of dirt are removed from the air stream. The job of either the screened inlet or pre-cleaner is to remove coarse particles and relieve the air filter of that much load.

After the air passes through the screened air inlet or pre-cleaner, it is drawn into the air cleaner through a tube called an air stack or duct.

(1) OIL-BATH AIR CLEANER. At the bottom of an oil-bath air cleaner is an oil cup (Figure 60). When the air reaches this point, its path of travel changes abruptly to an upward movement. Oil is picked up and carried along with the air. This action coats the dirt particles

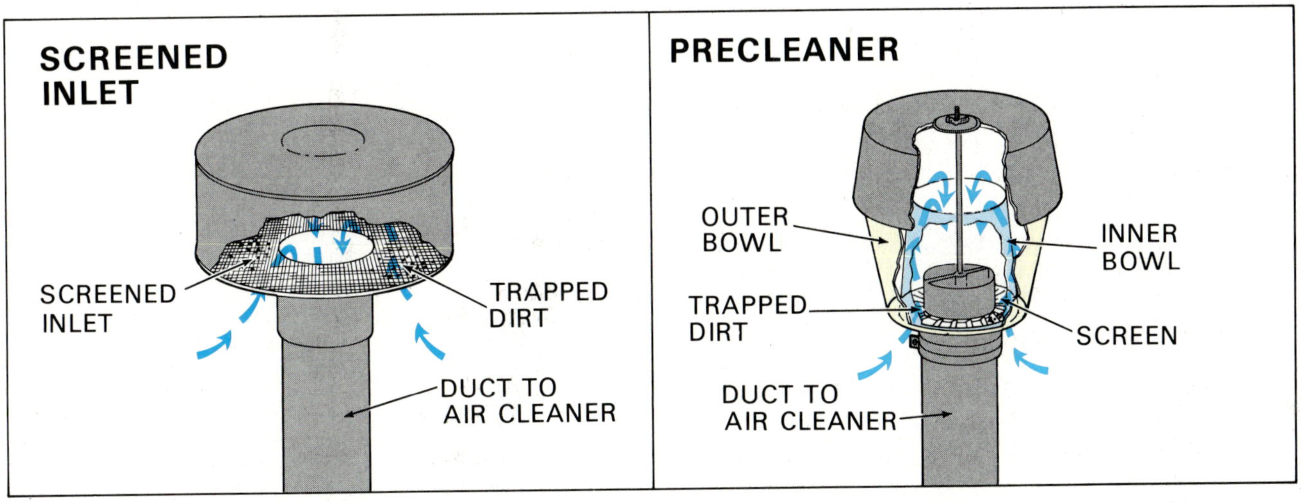

FIGURE 59. Tractor air cleaners may consist of a screened air inlet or a pre-cleaner that removes coarse particles.

FIGURE 61. Oil-bath air cleaners are always mounted upright (vertically) so the oil will remain in the cup at the bottom of the cleaner.

with oil and causes them to lodge in the mesh of the filtering element. As more oil-coated dirt collects in the filtering element, it drains back into the outer chamber of the oil cup. Here the dirt settles out, and the oil is used again to trap more dirt.

Oil-bath air cleaners are always mounted upright (Figure 61).

(2) DRY-PAPER FILTER AIR CLEANER. Air is cleaned in a dry-paper filter air cleaner by passing it through a special pleated-paper element which filters out small particles and holds them on the outer surface of the filter element (Figure 62).

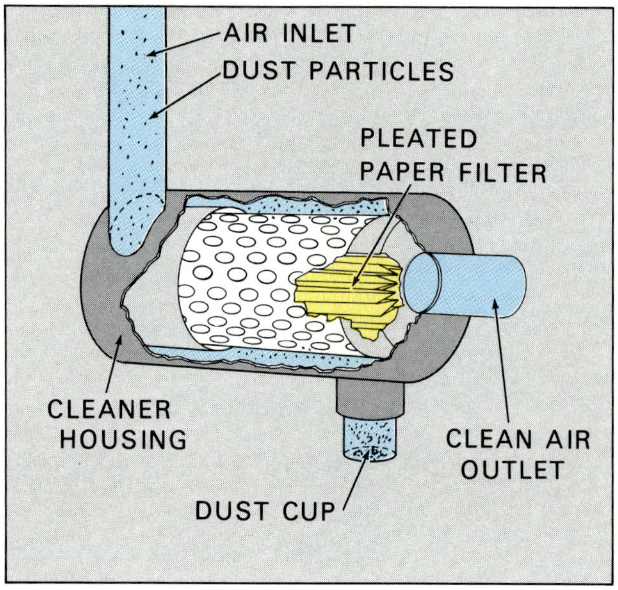

FIGURE 62. Principle of dry-paper filter air cleaners.

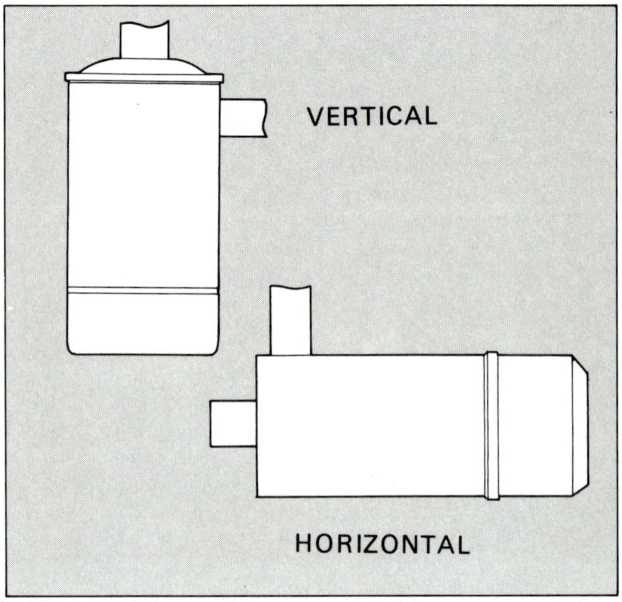

FIGURE 63. Dry-paper filter air cleaner mountings.

Dry paper filter air cleaners may be mounted either vertically or horizontally (Figure 63).

b. How to Service Air Cleaners

Under normal operating conditions, most manufacturers recommend that either type of air cleaner be checked for dirt accumulation each day. **Under very dirty conditions,** check every 5 hours.

Procedures are different for servicing each type of air cleaners. The two types discussed are:

(1) Oil-Bath Air Cleaners.
(2) Dry-Paper Filter Air Cleaners.

(1) OIL-BATH AIR CLEANERS. To service oil-bath cleaners, proceed as follows:

1. *Loosen clamp and remove the oil sediment cup from the lower part of the air cleaner (Figure 64).* When the dirt level in the cup is more than .625 cm (1/4-inch) deep, pour out the oil and sediment.
2. *Clean the cup with a solvent (Figure 65).*
3. *Check air-intake pipe (air stack) for dirt accumulation.*
   This will not need cleaning often, but it is good practice to check it each time the oil cup is serviced.
   If dirty, swab it out with a clean rag. The more dirt you allow to accumulate in the stack, the more it acts as a choke to air flow which increases fuel consumption.
4. *Refill oil cup with new oil to the oil level indicated on the cup (Figure 66).*
   Most operator's manuals recommend that you use the same weight oil as used in the engine

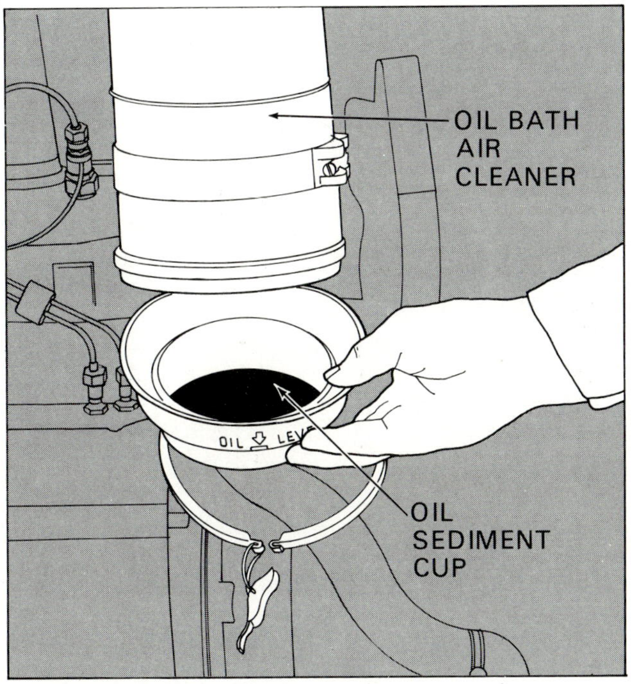

FIGURE 64. Removing the oil cup.

crankcase. If you use an **oil that is too heavy**, the washing action decreases. It also acts as a choke on the engine and increases fuel consumption.

Oil that is too light will be pulled into the engine, lowering the oil level in the cup until the cleaner is not efficient. With diesel engines, the oil will be used as fuel, causing the engine to run wild. When this happens, you have no control of engine speed.

A non-detergent oil may be used which will cause dirt to settle out better.

Do not use old crankcase oil. It may contain

FIGURE 65. Cleaning the oil cup.

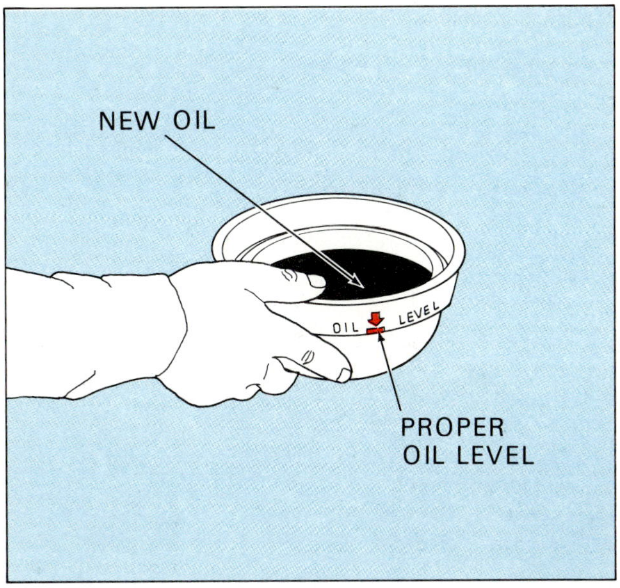

FIGURE 66. Refilling the oil cup.

unburned fuel which evaporates quickly and lowers the oil level in the cup. It also contains dirt which limits the washing action.

Do not overfill oil cup. On carburetor engines, overfilling increases fuel consumption and causes loss of power. On diesel engines, the oil may be drawn into the engine and used as fuel, causing the engine to run wild.

5. *Replace oil cup and tighten clamp.*
6. *Check air duct between cleaner and carburetor (or intake manifold on diesel engines) for holes or loose clamp connections (Figure 67).*

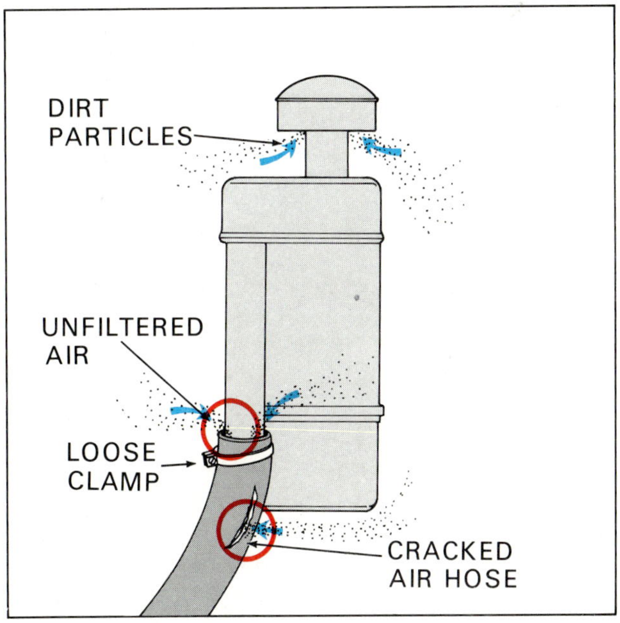

FIGURE 67. Loose clamps, or holes in air duct between air cleaner and carburetor, allow unfiltered air to enter. Dirt passes directly to engine, causing rapid wear.

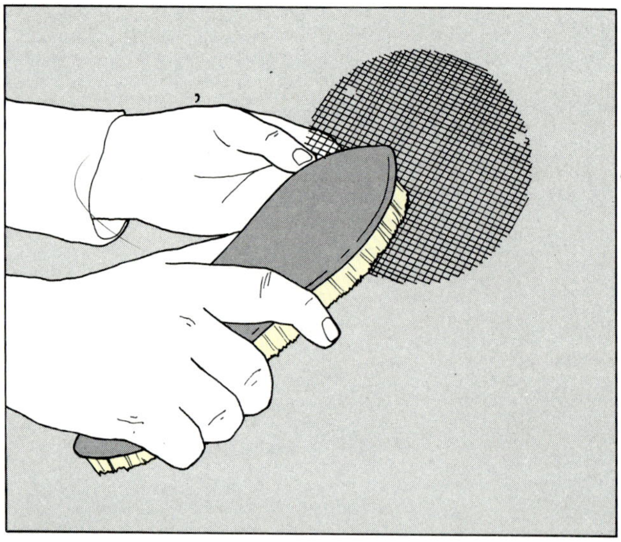

FIGURE 68. Clean screened air inlet with stiff brush. Clean pre-cleaner by removing bowl and emptying contents.

Air entering the engine in this manner bypasses the air cleaner. Dirt particles pass directly to the engine, causing rapid wear.

7. *Check screened air inlet, or pre-cleaner, and remove trash if necessary.*
   A stiff brush is good for cleaning the screen (Figure 68).

(2) DRY-PAPER FILTER AIR CLEANERS. To service dry-paper filter air cleaners, proceed as follows:

1. *Squeeze dust unloader or check indicator if supplied on your air cleaner (Figure 69).*
   The dust unloader is a rubber duck-bill arrangement that is held shut by low vacuum while the engine is running.

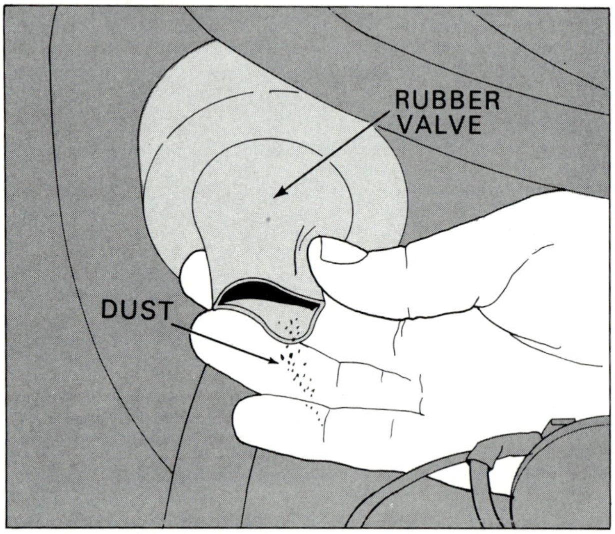

FIGURE 69. Some dry-paper filter air cleaners have a dust unloader. The rubber valve (rubber duck-bill arrangement) is squeezed to make certain it is open and free of obstruction.

FIGURE 70. Remove hood or grill to check filter element.

   Squeezing makes certain the valve is working and not blocked by an obstruction.
2. *Remove hood or grill, if necessary, to provide access to air cleaner (Figure 70).*
3. *Wipe off dust accumulated around end of cleaner where element will be removed.*
4. *Loosen hand screw or clamp that holds the end cup onto the end of the cleaner and remove cup and filter element (Figure 71).*
5. *Empty cup, wipe out dust, replace or clean filter element and replace cap.*
   (Procedures for replacing the filter element are given in the publication, TRACTOR MAINTENANCE—PRINCIPLES AND PROCEDURES, AAVIM).

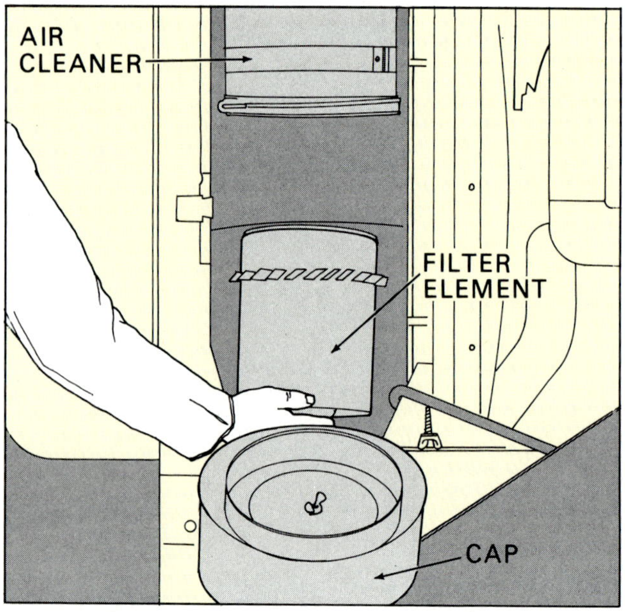

FIGURE 71. Removing the dry filter element.

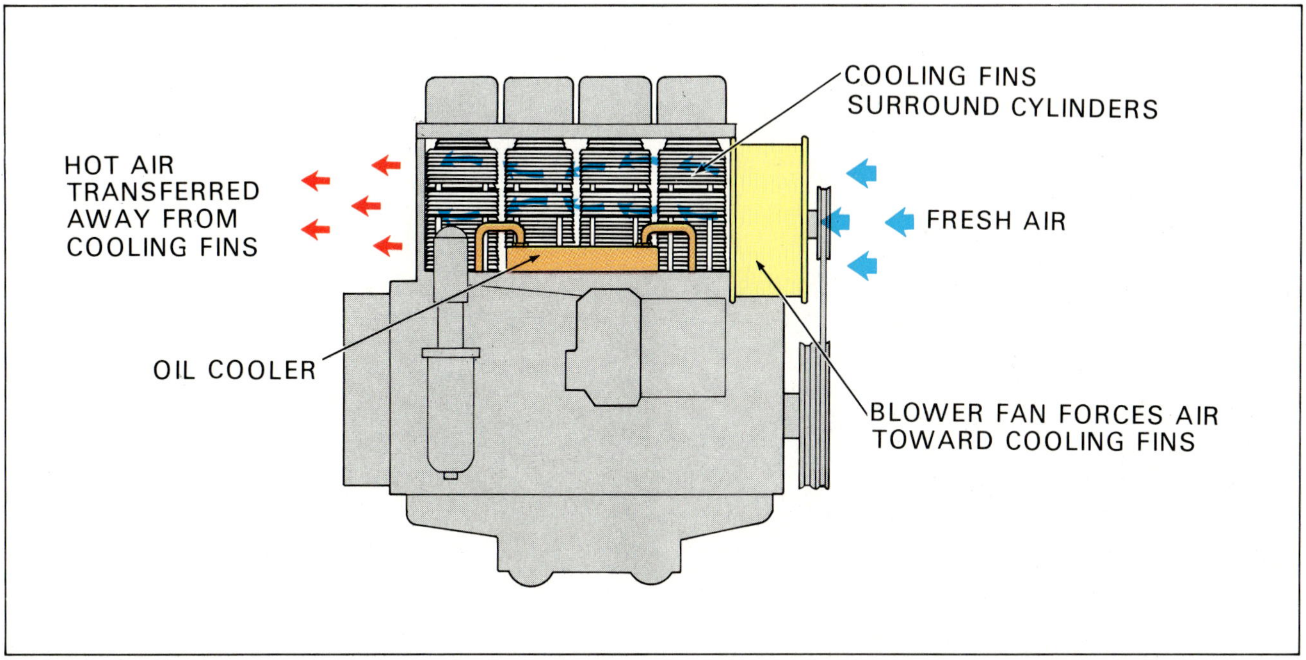

FIGURE 72. High velocity air is directed over the cooling fins to transfer heat away from the cylinders.

3. SERVICING THE COOLING SYSTEM

When fuel burns inside an engine, the temperature may momentarily reach 1765°C to 2400°C (3500° to 4000°F). Of course, that temperature exists for such a short time that it has no bad effect as long as excess heat can be easily removed through the tractor exhaust and cooling systems. Approximately 40 percent of the excess heat is given off through the cylinder walls. An adequate cooling system must be provided to get rid of the rest of it.

Tractors may have either an air-cooled or a liquid-cooled engine.

Procedures for servicing the cooling system are different and are explained under the following headings:

a. Air-Cooled Engines.
b. Liquid-Cooled Engines.

a. Air-Cooled Engines

The burning of fuels in the combustion chambers creates extremely high temperatures. In air cooled engines heat is transferred to the atmosphere through cooling fins located on the outside of the cylinder walls (Figure 72).

To service the cooling system of the air-cooled engines, proceed as follows:

1. *Lift the hood or remove the side panels.*
2. *Clean cooling fins of all dirt, grease and debris (Figure 73).* Use air hose. Solvents should not be used.
3. *Clean oil cooler if installed (Figure 74).* Use air hose.

FIGURE 73. Use an air hose when cleaning cooling fins.

FIGURE 74. Use air hose to clean oil cooler.

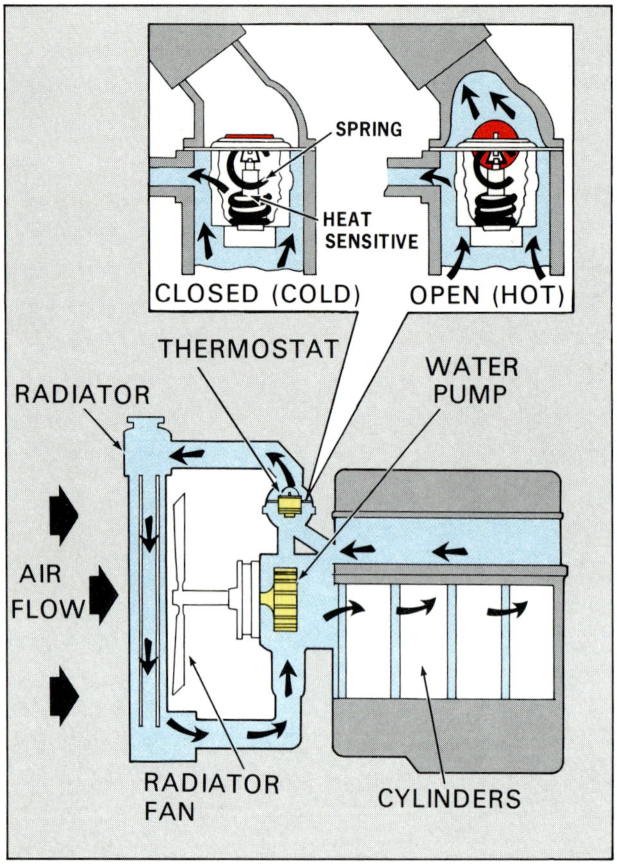

FIGURE 75. Liquid cooling system.

b. Liquid-Cooled Engines

As an operator, you need to understand how the cooling system works. You will then better appreciate the importance of proper care. Figure 75 shows how a **forced-circulation cooling system** works—the kind used on most present-day tractors.

The water pump pushes the heated coolant from inside the engine block, past the thermostat to the upper part of the radiator, then down through the radiator tubes. The fan, on the opposite end of the shaft from the pump, pulls air through the radiator tubes to help speed cooling. The coolant is then forced through the lower hose and pumped back into the engine water jacket. When the engine is cold, the thermostat closes and blocks circulation until the coolant temperature reaches about 71°C (160°F).

For the coolant, a mixture of water and antifreeze is used. When the engine is cold, the thermostat remains closed. The coolant passage is blocked, except on some systems in which there is provision for a small amount of coolant to circulate past the pump as shown in Figure 76. When the coolant reaches a temperature of about 71°C (160°F), the thermostat starts to open and continues until completely open at temperatures

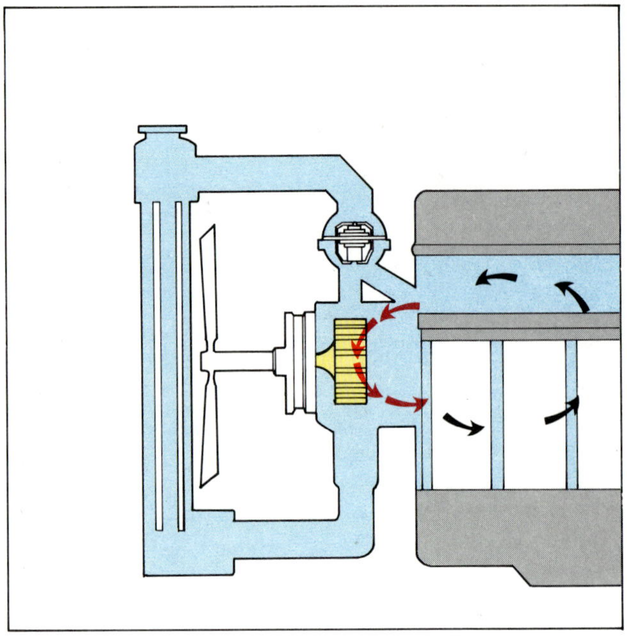

FIGURE 76. Some cooling systems allow a small amount of coolant to circulate past the pump.

of about 82°C to 91°C (180°F to 195°F).

A **pressure-type radiator cap** is used with forced-circulation cooling systems. This cap makes it possible for you to operate your tractor at higher temperatures.

Water at a temperature of 82°C to 91°C (180°F to 195°F) is so hot it tends to vaporize readily. Without a pressure cap, water vapor would be free to pass through the overflow tube which would probably cause you to have to refill the radiator several times a day. With a pressure cap, the vapor stays inside the radiator.

Actually, the pressure cap consists of two caps (Figure 77). The outer one covers the radiator opening, protecting it from dirt and providing a mounting for the inner cap. The inner cap fits inside the radiator neck and is held in place with a spring which applies a pressure of 27.5 kilopascals to 68.9 kilopascals (4 to 10 pounds per square inch). The spring acts as a **pressure-release valve.** If the pressure in the radiator gets above spring pressure, the cap permits air or vapor or both to pass into the upper section and out through the over-flow tube.

If the cooling system on your tractor is in good condition, the coolant probably won't get hot enough to lose vapor under normal operating conditions. With 27.5 kPa to 68.9 kPa (4 to 10 pounds pressure per square inch), the boiling point of water is changed from 100°C to about 110°C (212°F to about 230°F).

The pressure-release valve functions when you warm up a cold engine. If the coolant level is low, the heated air in the top of the radiator may build up enough pressure that some air will escape past the relief valve. Or, if you have completely filled the radiator

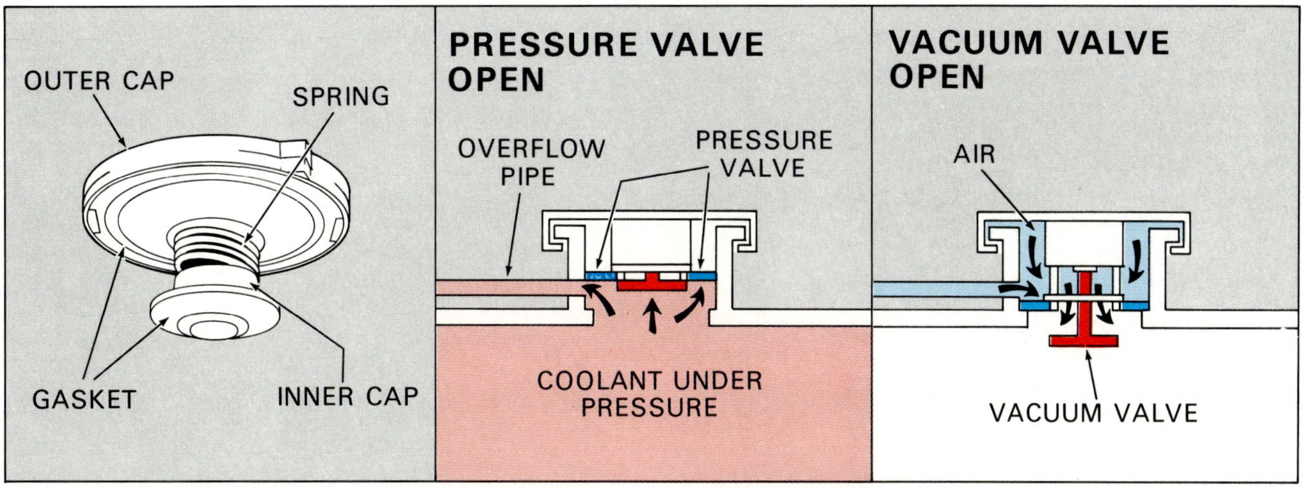

FIGURE 77. Pressure-type radiator cap. Unit consists of outer and inner caps. Inner cap provides pressure-release valve and a vacuum valve to allow air to enter when coolant cools.

so that little or no air space is left, the heated coolant will expand and some coolant will be released by the valve.

When a cooling system works under pressure, still another valve is required. It is a **vacuum valve.** It fits in the center of the pressure-release valve and allows air to enter the radiator when the coolant cools. Without it, a vacuum would develop when the engine cooled, causing the radiator hoses to collapse and air leaks to develop around the water-pump shaft and any weak spots in the radiator.

To service the **liquid-cooled system**, proceed as follows:

1. *Remove the radiator cap and check level of coolant.*
   If the coolant is more than 5 cm (2 in) below the neck of the radiator, it is in need of filling.

   CAUTION! If you have been running the engine long enough for it to become heated, remove the radiator cap slowly (Figure 78). Give it a slight turn and you will feel it partially release without coming off. In this position any excess pressure will be relieved. If you remove the cap completely, you may get sprayed with steam and hot water.

   If the pressure cap is worn so that there is no buildup of pressure in the radiator, replace it with a new one. It is best to **buy it from a dealer who sells your type of tractor** in order to be sure that the relief valve is the proper one for the right pressure setting. The wrong cap can cause overheating or cause the radiator to burst.
2. *Refill radiator with coolant to within about 5 cm (2 in) of radiator neck (Figure 79).*
   This amount of space is about average for the various makes of tractors. However, manufacturer's recommendations range from as little as 1.2 cm (½ in) to bottom of filler neck to 10 cm (4 in) when measured to the top of the filler neck. Because of this variation, check the recommendations in the operator's manual for your tractor.

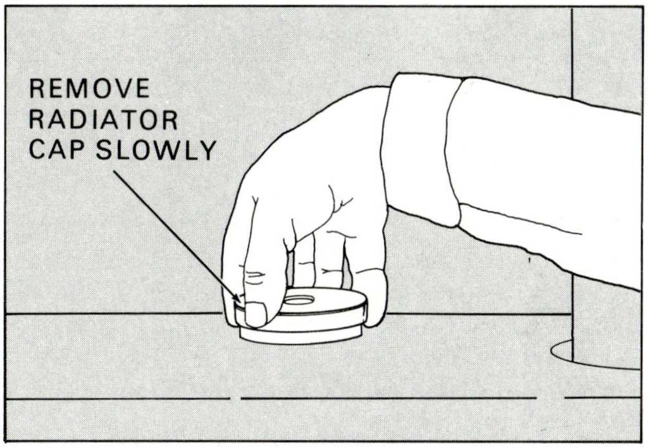

FIGURE 78. Remove radiator cap on a hot engine slowly and carefully.

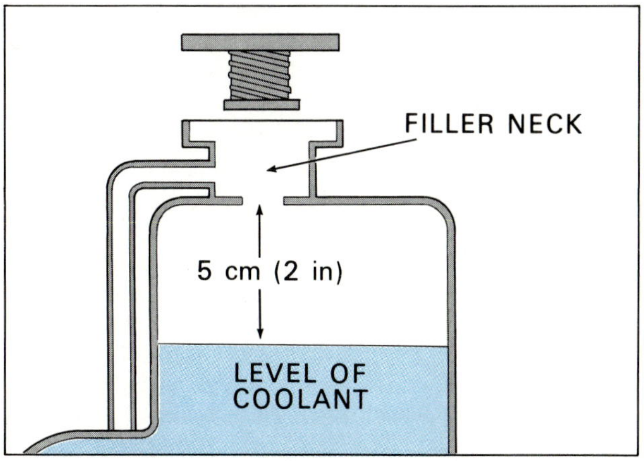

FIGURE 79. Proper coolant level in a radiator.

FIGURE 80. Use an air or water hose to remove trash from radiator.

Filling to this recommended level is important for two reasons:

- There is **room for the coolant to expand** without overflowing, which is particularly important if you are using antifreeze.
- The **coolant is exposed to a minimum of air**.

If you let the **coolant level get low**, there will be a large air pocket in the top of the radiator from which the coolant will absorb oxygen. Oxygen and moisture cause rust to form in the cooling system. The more air in the radiator, the faster rust forms, unless you use a **rust inhibitor**.

If you are using **antifreeze**, it is almost certain to contain a rust inhibitor. If you are using water, you can buy a can of rust inhibitor and add it to the water to stop rusting action. However, even with a rust inhibitor, you still need to keep the radiator filled to the proper level. The reason is that the rust inhibitor contains an **anti-oxidant** which takes up oxygen from the coolant and this controls rust as long as the anti-oxidant remains active. However, the more air the coolant is exposed to, the more rapidly the capacity of the anti-oxidant is used up. Use antifreeze for cold weather operation. Check your operator's manual for the proper mixture to protect against anticipated cold weather.

If you are using **water**, be sure it is clean. **Dirty water** clogs the cooling system and makes it less effective. Use soft water (rain water or softened water) if available. Hard water causes corrosion and deposits in the engine and radiator which slows the movement of heat, causing your engine to run too hot unless you clean and flush the system regularly.

It is best to **fill the radiator while the engine is still cold**. If you have to fill it after it is hot, idle the engine and pour in the coolant very slowly. Adding coolant rapidly to a hot engine can crack the head or engine block.

3. *Check for leaks around radiator, radiator hoses and hose clamps*.
4. *Remove trash collected on front of radiator or on radiator screen (Figure 80)*.

   If you have air or water under pressure, use a hose and direct the stream against the radiator from the side next to the engine. Be careful not to damage the fins in the radiator.

   There are some very trashy operating conditions, such as those involved in corn harvesting, when the radiator should be checked and trash removed by hand several times a day in the field.

FIGURE 81. Remove battery caps to check water level.

4. SERVICING THE BATTERY

While the regular interval for servicing the battery (or batteries) in your tractor is 25 hours, you should take a look at it daily. Proceed as follows:

1. *Check the electrolyte level of the battery by removing the caps (Figure 81).*
   Electrolyte is the acid solution in the battery. If the level is allowed to remain below the top of the plates, they will dry out and shorten the life of the battery. The level should be 10 mm (3/8 in) above plates.

   CAUTION! Do not smoke or light a match while caps are removed (Figure 82). If the battery has been charging, hydrogen gas is present. A concentration as low as 7 percent may burn or explode in the presence of a spark or flame and spatter acid

FIGURE 82. Never smoke or light a match near a battery while caps are removed.

FIGURE 83. Apply silicone to cables and terminals.

2. *If the electrolyte is below the correct level, add clean soft water to cover the plates to the bottom of the filler tube.*
   A battery will consume about one teaspoon of water per cell each month, more in extremely hot weather. If water is added in freezing weather, run the engine at least one hour to be sure the water and electrolyte are mixed thoroughly.
3. *Clean cables, terminals and supports and apply a coat of silicone spray for protection (Figure 83).*
   Check battery support brackets for corrosion and clean if necessary.
4. *Tighten battery connections.*
   Keep connections tight at all times. Loose cables will cause electric arcing, pitting of connections and eventual battery failure.

   Keep the positive battery post covered with a rubber boot on the end of the cable (negative post if a positive grounding system).

5. LUBRICATING THE CHASSIS

On most tractors there are a number of places which you should grease daily or after each 10 hours of operation. Regular greasing of these areas saves wear and protects the bearings. It also makes it easier for you to operate the tractor. Most of these fittings are on the hand- (or foot-) operated parts—the steering mechanism, brakes or clutch (Figure 84).

Some manufacturers have lengthened the service periods for lubricating these points on the chassis to as long as 50 hours. Other manufacturers are gradually replacing bearings that need daily servicing with prepacked sealed bearings that require no servicing (Figure 85).

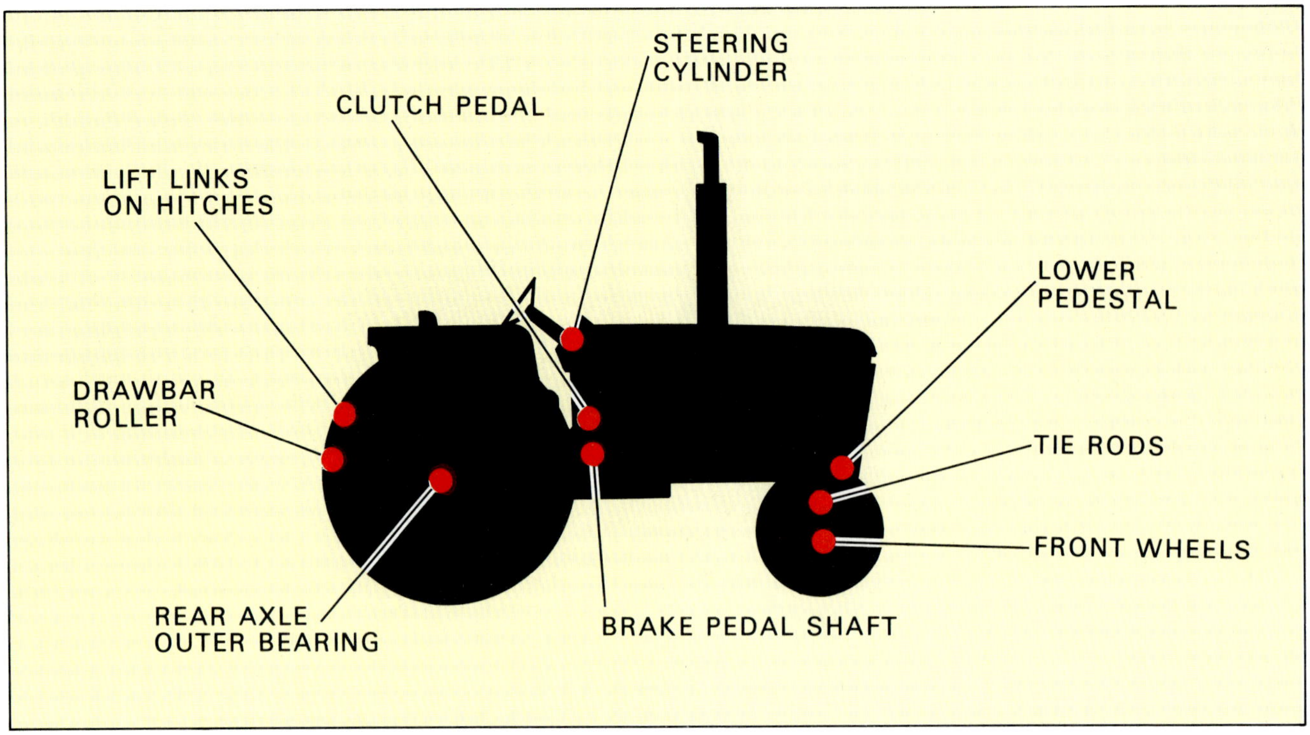

FIGURE 84. Examples of grease-fitting locations on tractors.

Use a check list to make sure you find all grease fittings. It is almost impossible to remember all of them. Your operator's manual is your best source of information.

Most operator's manuals, when mentioning the kind of grease you should use, refer to "chassis grease," "pressure-gun grease," "No. 1 pressure-gun grease," "ball-and-roller bearing grease" or "general-purpose grease." When you ask for any of these, your supplier will probably offer you two different greases: (1) a **multi-purpose grease** which is relatively new but is satisfactory for both chassis and water-pump use, or (2) the older type of **chassis grease** which is suitable for greasing the chassis but not the water pump. A special water-pump grease is needed with the latter if your tractor has a pump that requires greasing.

It is very important to keep the grease free of dirt and water. It usually becomes contaminated after the container is opened and while it is being stored between uses. To meet this problem, several means have been developed for keeping the container covered from the time it is opened until the supply is used. Figure 86 shows two of these methods.

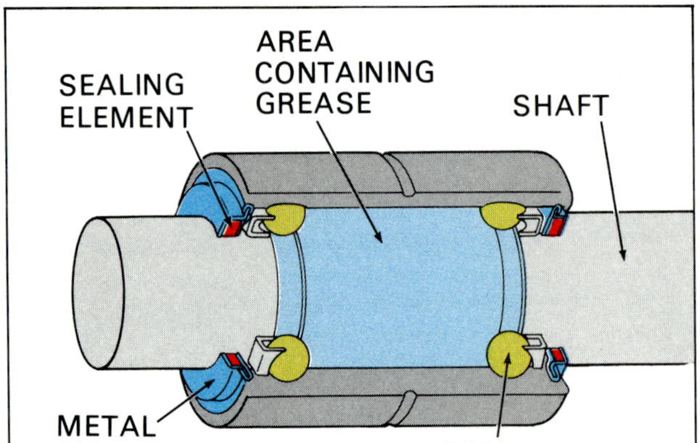

FIGURE 85. Prepacked ball bearings are sealed to retain grease and keep out dirt. They require no servicing.

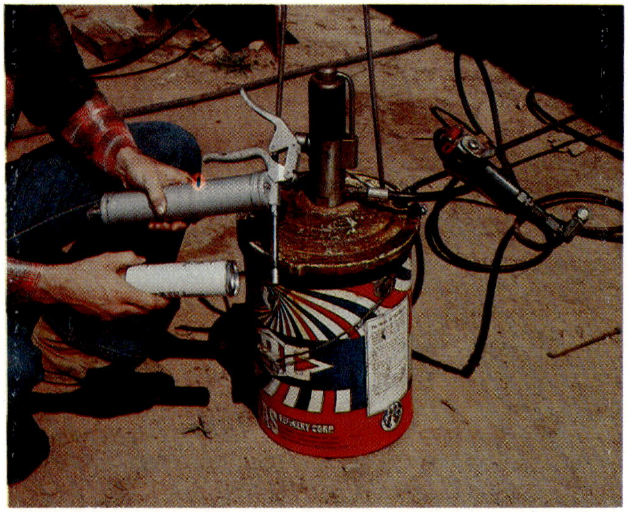

FIGURE 86. Two methods used to protect grease from dirt and water: (left) sealed cardboard grease cartridge, for hand or air-pressure guns, and (right) dispenser that attaches on top of grease pail which may be operated by hand, with compressed air or by an electric motor.

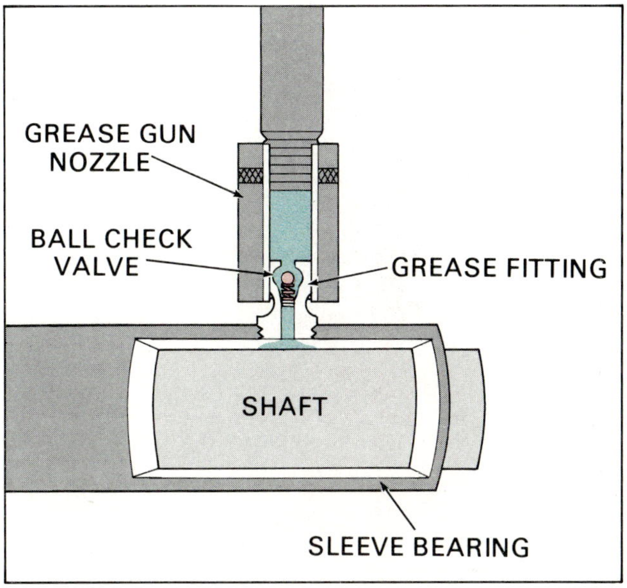

FIGURE 87. Pressure from gun forces grease past ball check valve to bearing surface. When gun is removed, the valve closes to keep out dirt.

FIGURE 89. Hold grease gun directly in line with fitting.

Grease guns are used to lubricate the chassis (Figure 86). They force the grease into the bearing through a grease fitting (Figure 87).

When using a **grease gun**, proceed as follows:

1. *Wipe end of grease fitting with a clean cloth (Figure 88).*
   This step is important. If you don't clean the fitting, any dirt accumulated on the end is forced into the bearing with the new charge of grease. This dirt makes bearings wear fast.
2. *Place gun on fitting and hold directly in line with it (Figure 89).*
3. *Pump gun slowly.*
   If the bearing has a **grease seal**, grease lightly. Too much pressure destroys the seal and allows dirt to enter. Check your operator's manual.

   If the bearing has **no seal**, apply lubricant until old grease is forced out and new lubricant appears at edges.

   Pressure of grease from the gun holds the gun coupler tightly against fitting.
4. *Remove gun by swinging through a 30° arc (Figure 90).*
5. *Wipe excess grease from fitting.*

FIGURE 88. Be sure grease fitting is clean before applying grease gun.

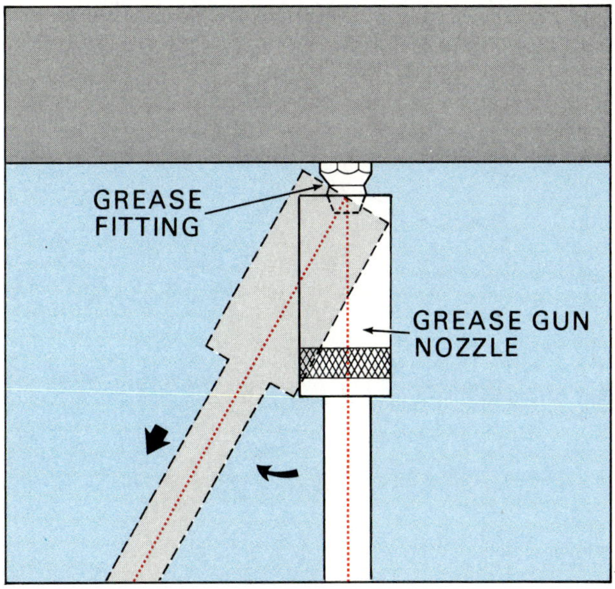

FIGURE 90. To remove grease gun, swing the gun away from the grease fitting in a 30° arc.

6. SERVICING THE DIESEL FUEL FILTER SYSTEM

Water and sediment have a bad effect on the diesel fuel injection system. **Sediment** grinds away the finely-fitted parts of the injection pump. **Water** causes the parts to rust. A very small amount of rust can cause serious damage. Water also interferes with the proper feeding of fuel and causes the engine to run rough.

Water is more of a problem in diesel fuel than it is in gasoline. Water separates from gasoline rather quickly, but the weight of diesel fuel is so close to the weight of water that they mix rapidly and separate slowly. When the tractor is motionless during the night, water has time to separate from the diesel fuel and settle to the bottom of the tank. The trapped water should be removed from the fuel tank each morning.

Tractor manufacturers are aware of this situation and have provided for removing the water by such means as shown in Figure 91.

To service the diesel fuel filter system, use the directions in your operator's manual for the make and type of your tractor. If you do not have a manual, use the following procedures.

1. *Check to remove water and sediment at one of the three locations shown in Figure 91.*
   (a) If equipped with drain cock, proceed as follows (Figure 92).
      (1) Open and drain about .5 liter (1 pint) of fuel (to make certain water and sediment are removed).
      (2) Close drain cock.
   (b) If equipped with water trap and sediment bowl and if sediment bowl is more than ⅓ full of water, proceed as follows (Figure 93).
      (1) *Close fuel valve under fuel tank.*
      (2) *Loosen thumb nut under bowl and swing bail to one side.*
      (3) *Remove glass bowl, empty and clean.* Be careful not to lose the gasket the filter bowl presses against. When bowl is removed, the gasket may be on top of bowl or may remain in top of filter assembly.
      (4) *Wash the glass bowl in kerosene to remove dirt and sediment.*
      (5) *Install bowl loosely.* When installing the bowl, be sure to check the gasket. If it is not in good

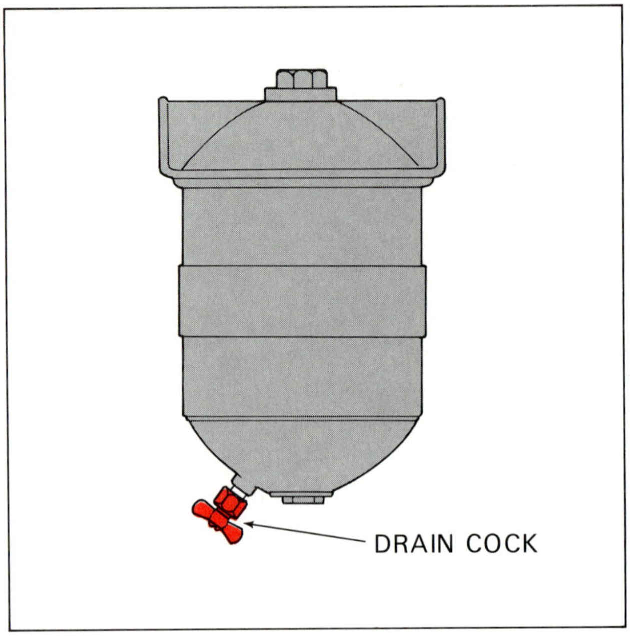

FIGURE 92. Fuel filter equipped with drain cock.

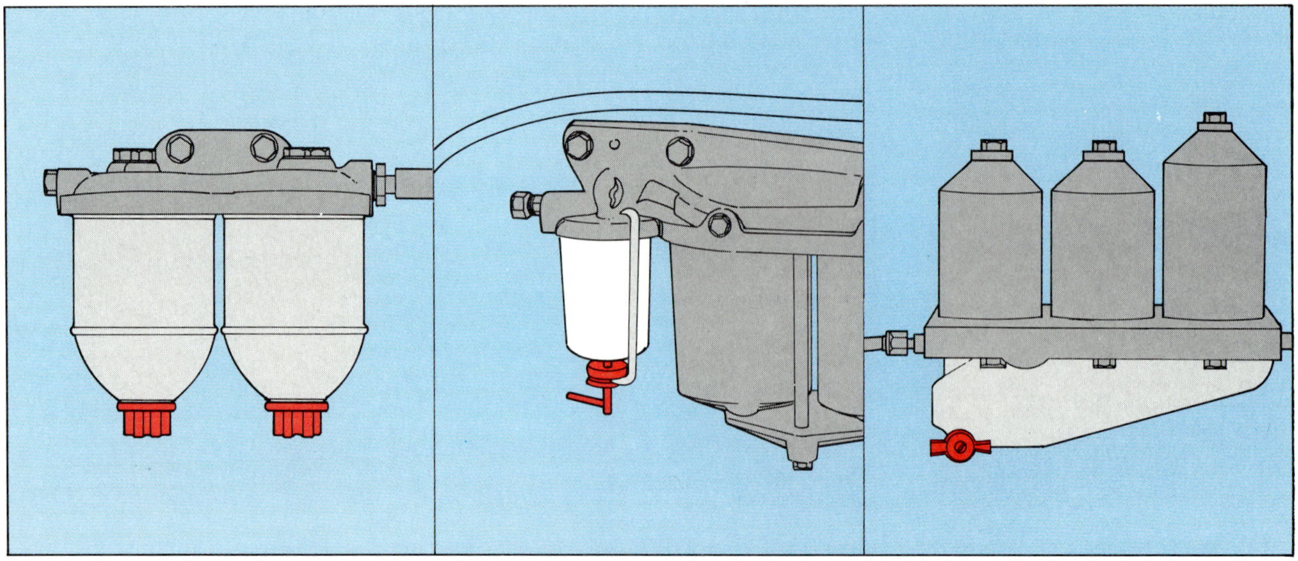

FIGURE 91. Arrangements for daily removal of water and sediment from diesel fuel system.

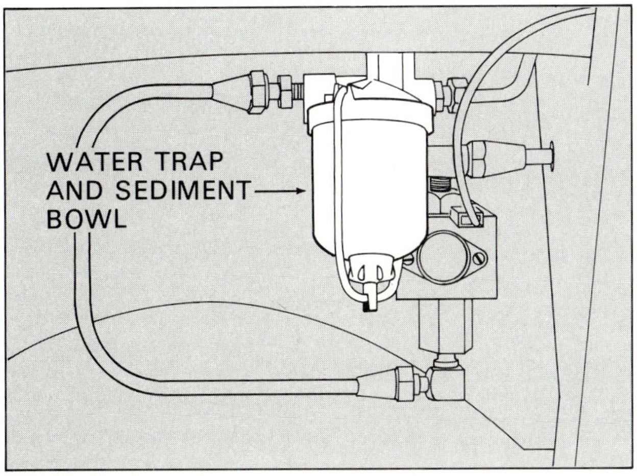

FIGURE 93. Fuel filter equipped with water trap and sediment bowl.

FIGURE 94. Check fuel level.

shape, replace it with a new one.

(6) *Open fuel valve and allow bowl to fill.* With the bowl mounted loosely, the fuel forces air out of the top of the bowl.

(7) *Tighten thumb nut under bowl when air is completely removed.*

7. REFUELING THE TRACTOR

You should check the fuel level before starting the engine (Figure 94), especially if someone else operated the tractor previously. Running out of any type of fuel is inconvenient. However, if you run out of diesel fuel, you have to bleed the system to remove air before restarting.

Proper refueling of your tractor is more than just filling the tank with fuel. The time and method of refueling are important because of their influence on the following:

- **Keeping out moisture.**
- **Keeping out dirt.**
- **Avoiding fire hazards.**

Most operator's manuals suggest that you fill the fuel tank at the end of the day or whenever you finish using the tractor. The reason is to keep **moisture** from accumulating in the tank (Figure 95).

Most water in fuel comes from moisture condensing inside the fuel tank. This is true of all fuels except LP-Gas. Here is why. As fuel is used, air is drawn in. Much of the time the air contains considerable moisture. The warmer the air the more moisture it can hold. At the end of the day, the fuel tank is full of warm moist air which, when cooled at right, deposits moisture on the sides of the tank. As it collects, it runs into and mixes

FIGURE 95. How a fuel tank collects moisture. Warm moist air replaces gasoline or diesel fuel used from fuel tank. As air cools, moisture deposits on side of tank and runs into fuel. If fuel tank is filled immediately after tractor is used, there is no room for moist air.

with whatever fuel is left in the tank. (Because LP-Gas is always under pressure, no outside air enters an LP-Gas tank.)

Water doesn't do much damage to the tank. Most tanks have a protective coating that helps prevent rust. However, in a **carburetor engine** any moisture that gets past the sediment bowl in the feed line from the tank moves into the fuel pump (if the tractor has one) and then into the carburetor. Any sulfur present in the fuel combines with water and attacks the metal parts. In cold weather, water that has collected in the sediment bowl or fuel line may freeze and cause starting trouble.

Diesel fuel and water mix rather easily, and the water is slow to settle out—much slower than with gasoline. Yet if water gets to the injection pump, it causes galling and sticking of the finely-machined pump parts. The next step is an expensive repair job.

By filling the tank immediately after the tractor is used, there is enough time during the night for most of the water to collect at the bottom of the tank. This is the reason for the recommendation that water be drained from the tank before starting the engine.

In **diesel engines**, water speeds the wear of the close-fitting plungers in the injector pump. Water may also clog injector nozzles.

Fire hazards exist in handling any kind of fuel. But with gasoline and LP-Gas the risk is somewhat greater. There are two reasons: (1) both fuels vaporize at temperatures around −18°C (0°F)—only a small amount mixed with air (6 percent or less of gasoline vapor) will explode with a spark or flame, and (2) since the vapors are heavier than air, they settle to a low spot and remain there unless ventilated out.

Kerosene and fuel oil don't vaporize readily until the temperature reaches 44°C (110°F) or more. They are less hazardous than gasoline unless you spill them on a hot manifold or other hot surfaces. Then they vaporize readily. With so much vapor present, a spark or flame will readily ignite them.

Consequently, if your tractor is hot and you haven't time to wait 4 or 5 minutes for it to cool, wait and refuel the next morning. The risk of water in your tractor fuel becomes less important than the risk of fire or explosion from refueling a hot engine.

Procedures for refueling are discussed for gasoline and diesel fuels only. Since there are only such a small number of tractors using LP-Gas still in operation, refueling with LP-Gas is omitted. Consult your operator's manual for instructions on how to refuel an LP-Gas tractor.

To refuel with gasoline and diesel fuels, proceed as follows:

1. *Check against fire hazards.*
   Shut off engine and turn off all electrical switches to lessen the danger of a spark.

FIGURE 96. Smoking while refueling is dangerous.

   Allow engine to cool if extra hot. Be sure there are no open flames and that no one is smoking (Figure 96).
   Do not refuel in a closed building. Be sure you have a charged dry-chemical type fire extinguisher nearby.
2. *Check mouth of container or hose nozzle for dirt.* Remove any dirt with a clean cloth. Be sure to use closed containers if you are hauling fuel to the field. There is less chance of dirt and moisture entering the fuel tank if you fill your tractor directly from an overhead supply tank or pump (Figure 97).

FIGURE 97. Overhead supply tank.

FIGURE 98. Authorized fuel containers.

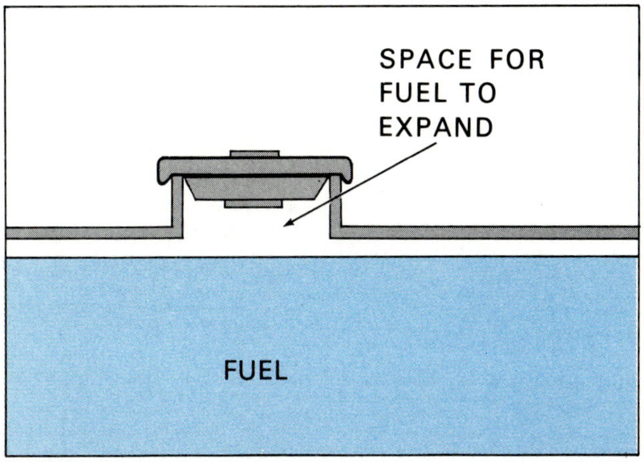

FIGURE 99. Leave space in neck for fuel to expand.

According to the National Fire Protection Association, all containers of flammable liquids must be properly labeled. While gasoline containers are traditionally painted red (Figure 98), there are no official colors for fuel containers.

3. *Remove dust and loose dirt from fuel tank cap and area around tank opening.*
4. *Fill fuel tank almost full (Figure 99).*
   Leave space in neck of opening for fuel to expand if heated. This precaution helps to avoid overflow and the possibility of fire.

   When filling from fuel drums, don't move them around any more than necessary. Unnecessary agitation of the fuel will cause water and dirt to mix with it.

   With diesel fuel, leave 7½ to 10 cm (3 or 4 in) of fuel in the bottom of the storage tank as a safeguard against moisture and dirt.

   If your **fuel supply is low** in the supply tank, you run a risk of drawing off both dirt and water into your tractor fuel tank. Rather than take this risk, line a clean funnel with a chamois and filter the fuel through it. The chamois removes both dirt and water.

5. *Replace fuel tank cap tightly.*
   Make it a practice to check the opening in the fuel tank occasionally. If it becomes clogged, a vacuum tends to develop in the fuel tank and fuel won't feed into the carburetor satisfactorily.

NOTES

III. Making Operational Adjustments on the Tractor

Procedures for making operational adjustments are given under the following headings:

A. Starting and Stopping Tractor Engines.
B. Preparing Tractors for Operation.

A. Starting and Stopping Tractor Engines

To start a tractor engine, you need to know more than which levers to pull or push, which valves to open or which buttons to press. You need to know such things as these: what makes an engine run, what safety precautions to take in starting an engine, how to protect the battery and starter from overloading, and how to warm up an engine properly to keep down rapid wear.

In addition, you must know how to stop the engine if something goes wrong.

While no two tractor models are alike, each instrument and control has a specific purpose. Your operator's manual is the best source of information for your tractor.

In the discussion thus far, almost all of the jobs apply to any make or model of tractor. But in starting a tractor engine you need to know whether it is a **carburetor-type** or a **diesel-type** engine.

Figure 100 shows two types of engines and the fuels

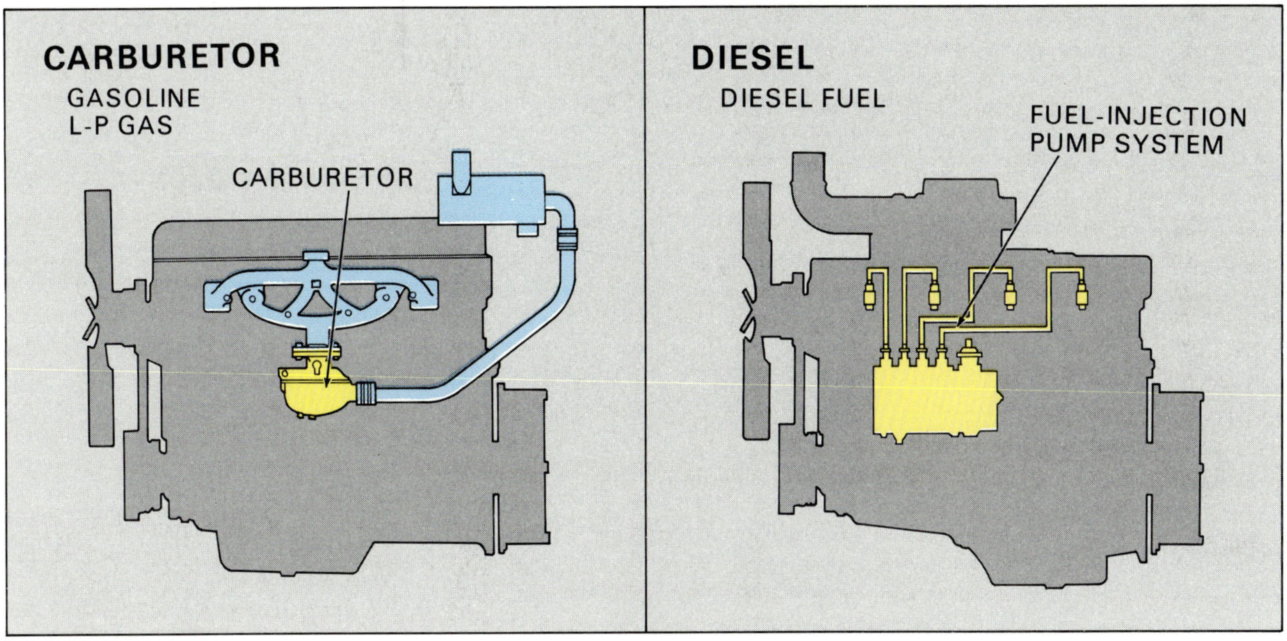

FIGURE 100. Two types of tractor engines: carburetor and diesel.

that have been developed for their use. Note that the two engines are similar in appearance. The difference is in the way fuel is supplied to the cylinders. The carburetor type engine in Figure 100 has a carburetor where fuel and air are mixed and drawn into the cylinder through an intake manifold. Not shown is the spark-ignition system that is used with this type of engine. It includes a coil, distributor, sparkplugs and wiring.

The diesel engine in Figure 100 has a pump mounted on the side instead of a carburetor. The pump **forces** fuel through separate tubes to each cylinder.

The discussion that follows gives this and other information along with reasons for the procedures that are recommended.

Although the starting procedures are similar for either carburetor- or diesel-type engines, the sequence varies slightly. For that reason, the starting procedures for each are given separately under the following headings:

1. Starting and Stopping the Carburetor Engine.
2. Starting and Stopping the Diesel Engine.

From your study of this section, you will be able to **describe the controls** and **list the procedures for starting and stopping tractor engines**.

1. STARTING AND STOPPING THE CARBURETOR ENGINE

Procedures are given as follows:

a. Starting the Carburetor Engine.
b. Stopping the Carburetor Engine.

a. Starting the Carburetor Engine

Proceed as follows:

1. *Check tractor maintenance record.*
   Perform maintenance as directed by the operator's manual. There are certain preventive maintenance jobs that should be done at intervals of 10, 50, 100, 500 and 1000 hours of operation.
2. *Perform preventive maintenance jobs.*
   The 10-hour (daily) service jobs are described in this manual. Procedures for performing other maintenance are given in TRACTOR MAINTENANCE, AAVIM.
3. *Clear the area.*
   Move obstacles from beneath and around the tractor. See that persons and animals are clear. Observe stationary objects that might be in the path of the tractor.
4. *Turn on fuel supply from tank if installed (Figure 101).*
   Before mounting the tractor, turn on the fuel supply. If there is no hand-operated cutoff valve, the fuel supply valve is opened automatically by oil pressure when the starter starts turning the engine. (If your tractor is equipped with a fuel pump, it may not have a fuel cutoff valve.)

   On most tractors, the fuel valve is opened completely to avoid leakage around the stem.
5. *Mount tractor.*
   Never attempt to start the engine from any place except the tractor seat.
6. *Check to see that shift lever is in neutral or park (Figure 102).*

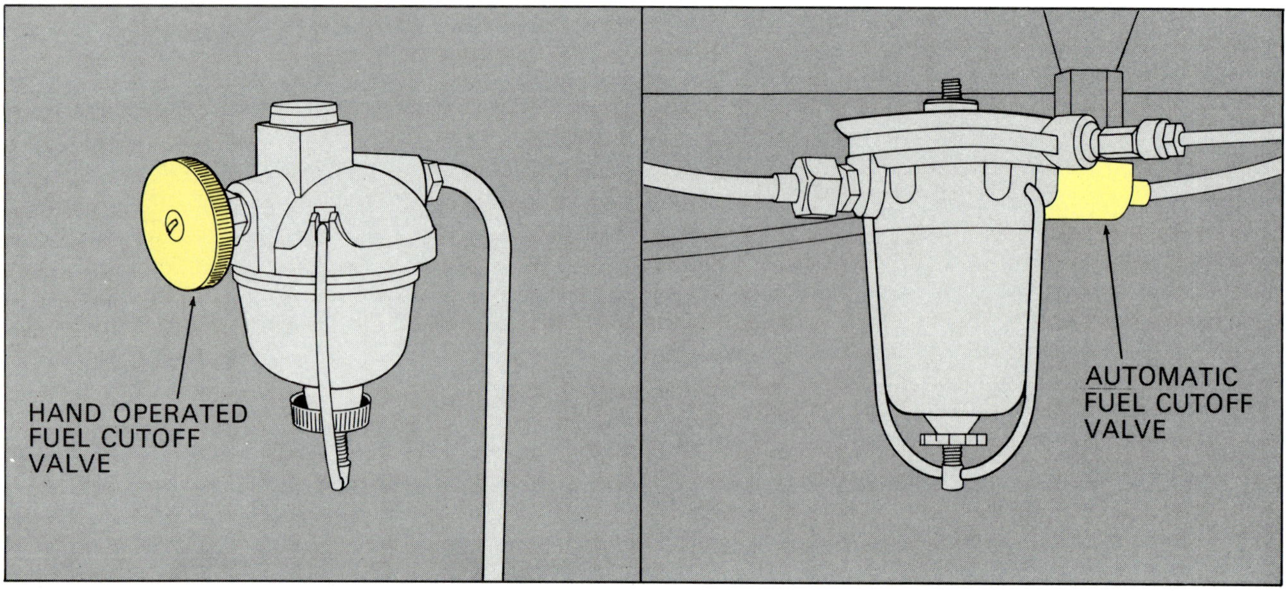

FIGURE 101. Types of fuel cutoff valves.

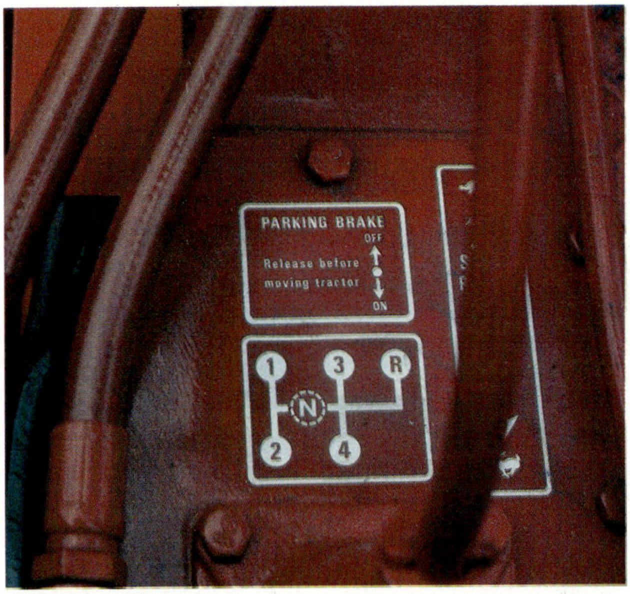

FIGURE 102. Make sure the shift lever is in neutral or park before starting.

Most tractors are designed so that the starter can't be engaged unless the shift lever is in neutral or park.

7. *Set speed-control (throttle) lever to desired position* (Figure 103).
For carburetor-type engines, set lever about ¼ to ⅓ open. The types of speed-control levers and their locations on tractors vary with different makes and models. In general, most speed control levers are located on the steering column, control console or the dash (Figure 103). On some tractors, however, a foot throttle control is also available (Figure 104).

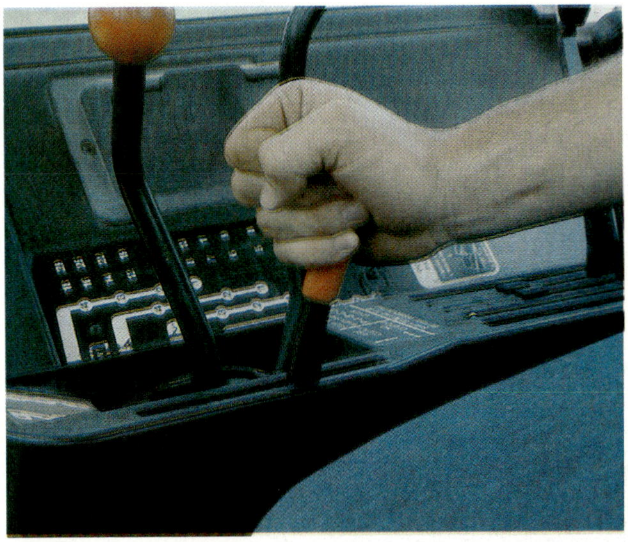

FIGURE 103. Setting speed-control (throttle) lever for starting.

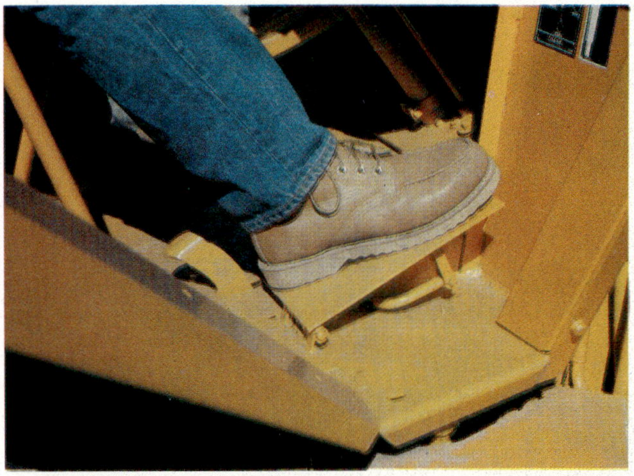

FIGURE 104. A foot throttle control.

With either type, always avoid sudden surges of acceleration and deceleration.

8. *Disengage clutch or depress inching pedal (Figure 105).*
By disengaging the main clutch, you relieve the drag on the starter caused by cold oil in the gear case and the drag caused by the gearing to the power take-off shaft or belt pulley. On most tractors, the clutch is foot-operated, as shown in Figure 105. Press the clutch down as far as it will go. Tractors with hand clutches are disengaged by pulling the lever back. These are often found on older-model tractors.
The inching pedal is used with power-shift transmissions. It is located in about the same position on the tractor as the foot clutch with hand-shift transmissions.

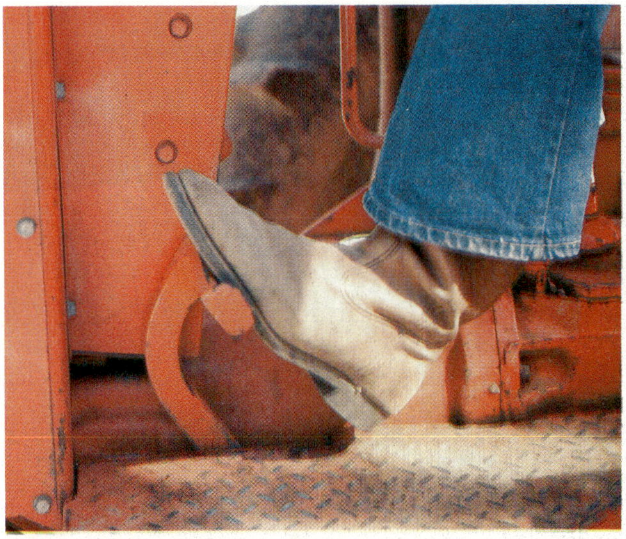

FIGURE 105. A foot operated clutch control.

FIGURE 106. Choke control is pulled out to start a cold engine.

9. *Set choke at desired position (Figure 106).*
   The choke control is usually pulled out all of the way in order to provide a rich gasoline mixture for easy starting of a cold engine. If the engine has been running and is warm, don't use the choke—open the throttle more.
10. *Turn ignition switch to "on" position (Figure 107).*
   The ignition switch on some tractors is combined with the light control switch. With these switches, the pointer points towards "I" (ignition) for the "on" position.

   Separate ignition switches provide simple "on" and "off" positions.

   Tractors equipped with magnetos usually have a push-pull switch that short-circuits (grounds) the magneto and stops the engine when the switch is pushed in. When pulled out, it breaks the ground so the engine can be started.
11. *Crank engine.*
   Most tractors are equipped with electric starters. The starter control may be one of the following:

FIGURE 107. Ignition switch.

FIGURE 108. Most oil gages indicate the proper working pressure.

- A **button** on the instrument panel or transmission case.
- A **combination** ignition and starting key.
- A **position** for the gear shift lever that engages the starter.

If your engine fails to start on the first try, wait until the engine and starter have stopped rotating before engaging the starter again. If you try to re-engage the starter while it or the engine is still rotating, there is danger of damaging the starter and the ring gear of the engine. **Don't operate the starter longer than 15 to 20 seconds**. Running the starter continuously for longer than this period is hard on the battery, causes the starter to overheat and can cause the starter switch to heat and fuse together.

Tractors with **power-shift transmissions** cannot be towed to start the engine. The transmissions in these tractors are made up of a system of planetary units that are controlled hydraulically. When the tractor engine is not running, there is no hydraulic control of the planetary units—power cannot be delivered from the rear wheels to the engine.

If you use jumper cables, connect the ground terminals first and disconnect the ground terminals first in order to avoid arcing.

12. *After engine starts, release foot clutch slowly.*
13. *Check oil-pressure gage (Figure 108) or warning light.*
   In cool or cold weather the gage will show high pressure at the start. As the engine warms, the pressure will gradually lower to its working pressure.

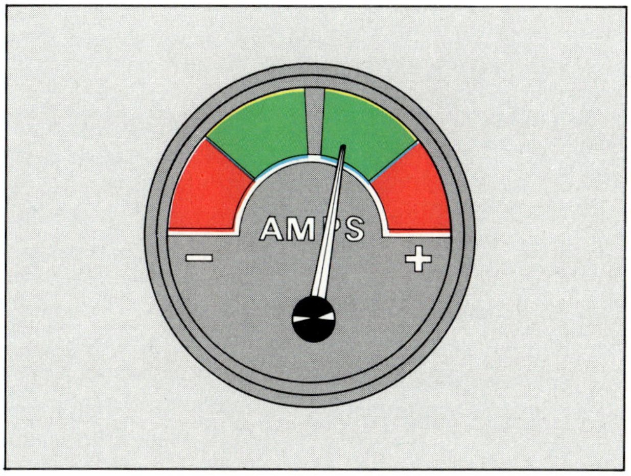

FIGURE 109. Ammeter should register on "charge" side during engine warmup.

FIGURE 110. The heat indicator shows when your tractor engine has reached the correct operating temperature.

(Some tractors are now equipped with a red warning light on the instrument panel in place of a gage. If it glows red, stop the engine.)

Most oil gages are located on the instrument panel. If it is not located here on your tractor, it is probably alongside the engine.

14. *Adjust choke for smooth operation.*
As the engine warms, push in the choke control gradually. If you pull it out too far or push it in too fast, the engine will miss.
15. *Adjust throttle to moderately fast idling speed.*
In general, a moderately fast speed for **carburetor** engines is 1/4 to 1/3 throttle.
This speed may seem too fast for a cold engine, but the extra speed makes oil available to bearings more quickly than a slow idle. Since the bearings have a very thin oil layer separating the moving parts when the engine first starts, it is important that they have a supply of oils as quickly as possible to keep down wear. At low temperatures, only a small increase in oil temperature will greatly improve its ability to flow.
16. *Check ammeter or warning light to see if battery is charging (Figure 109).*
If the ammeter shows discharge while the engine is warming up, check for the cause of trouble. (If warning light—labeled "GEN"—glows, this indicates discharge.)
17. *After one or two minutes, adjust to higher idling speed.*
Carburetor engines adjust to 1/2 to 3/4 throttle, or 1000 to 1200 rpm.
You increase engine speed to lessen warm-up time. Studies show that with a water-jacket temperature of 60°, cylinder wear is about 4 times greater than when the water-jacket temperature is 100° and is 16 times greater than at 150°.
18. *Watch indicator to determine when engine is warmed enough (Figure 110).*
Allow the engine time to warm up properly. You are not only saving considerable engine wear, but your tractor will give you much more power with less fuel when it has reached the proper operating temperature.

b. Stopping the Carburetor Engine

Procedures for stopping carburetor engines are given as follows:

1. *If engine has been heavily loaded, idle at about half throttle for two or three minutes.*
Unless the engine is idled, many high-compression engines will continue to run after the ignition is shut off or they may backfire and reverse direction. These problems are caused by spontaneous (self-generated) ignition by glowing carbon.
2. *Retard throttle and turn off ignition switch.*
Don't choke the engine to stop it. Choking leaves the cylinders with excess fuel which washes the cylinder walls as it passes around the piston rings and into the oil causing dilution.
3. *If the tractor is to remain idle overnight or longer, shut off fuel valve under tank.*
If your tractor has a vertical exhaust pipe, it is good to cover the end with a can to keep out rain. It will be much easier to start next time. Some have caps that automatically close when the engine is stopped.

2. STARTING AND STOPPING THE DIESEL ENGINE

Procedures are given under the following headings:

a. Starting the Diesel Engine.

b. Stopping the Diesel Engine.

a. Starting the Diesel Engine

Proceed as follows:

1. *Follow steps 1 through 8 under "Starting the Carburetor Engines."*
2. *Turn on fuel supply at fuel-stop control (Figure 111), but don't open throttle.*

   The fuel-stop control is a means of shutting off the fuel supply when stopping a diesel engine. On **some tractors the fuel-stop control is part of the throttle control**. By tripping a latch, the throttle lever can be extended past idling position for shutting off the fuel supply. With this type, you automatically turn the fuel supply on when you set the throttle for idling.

   On other tractors the fuel-stop control may be a separate knob or handle. If the control was pulled out and left the last time the tractor was stopped, it must be pushed in before fuel can be supplied to start the engine.

   This control operates independently of the fuel valve under the tank.

   Don't confuse the fuel-stop control with the compression-release lever. The latter is used to open the exhaust valves on all of the cylinders so that the engine may be turned over by hand when making timing or tappet adjustments. You can tell if the compression-release lever has released compression because of the ease with which the starter turns the engine.

FIGURE 111. A fuel-stop control.

NOTE! If tractor has not been operated for several weeks, turn the engine over a few times with the starter before opening the fuel stop. This lubricates the engine.

3. *Turn starter switch to "on" position.*

   The starter switch on a diesel differs from the ignition switch on the carburetor tractor. It simply energizes the circuit to the starter so you can use the starter whenever you are ready. When the switch is "off," the starter will not operate, but your engine will continue running if it is already started. However, unless indicated otherwise, you should leave the switch on at all times while the tractor is running.
4. *For cold-weather starting (below freezing), use provision on your tractor for easy starting.*

   Numerous devices and methods have been developed for starting a diesel tractor in below-freezing weather (Figure 112). They are ether, hand primer and pre-heater.

   Your operator's manual will tell you which method is used on your tractor. If you have no manual, you may be able to tell by examining the tractor.

   If ether is used, be sure to check for special operating instructions for the particular tractor you are starting.

   Do not add gasoline to diesel fuel in hopes of making the engine easier to start. Gasoline has no **lubricating qualities** so there is a chance the injectors will be damaged, and the **auto-ignition temperature of gasoline is higher** than for diesel fuel which makes starting more difficult.
5. *Depress clutch to ease starting load.*
6. *Crank engine several revolutions; then open throttle from ½ to completely open while engine is rotating.*

   The starter is usually controlled by a switch button or a key-operated switch. Limit starter operation to 15 or 20 seconds.

   Don't open throttle until starter has engine rotating. This keeps fuel from entering the cylinders too soon which can cause the engine to "kick back" and damage the starter drive mechanism.
7. *Follow procedures 12 through 18 under "I. a. Starting the Carburetor Engine."*

FIGURE 112. Methods used for helping start diesel tractors in cold weather.

b. Stopping the Diesel Engine

The following instructions apply to diesel engines with the exception of those that also operate as gasoline engines. (See operator's manual for latter type.)

1. *If the engine is operated on heavy load, let it idle at half throttle or faster for three or four minutes.*
2. *Completely retard throttle.*
3. *Turn off fuel supply (Figure 113).*
   This step may be accomplished by various means: by pulling a fuel-stop control handle, by tripping a throttle latch to allow the throttle to close beyond its lowest idling position, by moving a fuel shut-off knob to the "closed" position or by turning an electric fuel shut-off switch. The latter is usually a key switch which can be removed to prevent others from starting the engine.

FIGURE 113. One method of stopping the diesel engine.

B. Preparing Tractors for Operation

Before field operation, some additional jobs are necessary to make your tractor ready.

On completion of this section, you will be able to **give procedures for adjusting the seat, for adjusting the tread width, for adding weights** and **for adjusting tire pressure.**

They are discussed under the following headings:

1. Adjusting the Seat.
2. Adjusting the Tread Width.
3. Adding Weight.
4. Adjusting Tire Pressure.

1. ADJUSTING THE SEAT

Checking and adjusting the seat on your tractor are important for two reasons: to provide maximum comfort for the operator and for safer operation. Research has shown that operator fatigue (becoming tired) is one of the contributing factors to accidents. The reason for fatigue is that the vibrations absorbed by the body while driving a tractor are much faster (higher frequency) than vibrations caused from running or walking—the ones to which our bodies are best adapted.

If you have to strain to reach levers and foot pedals, you will tire faster and be unable to control the tractor properly. You will greatly increase your chance of having an accident. (Figure 114).

It follows then that a seat adjustment which provides for the most comfort and relaxation also provides the most freedom from fatigue.

FIGURE 114. Example of tractor seat out of adjustment.

When there is a change of drivers, the seat position should be rechecked and adjusted if necessary. To make certain it is, add it to your check list of jobs to do before starting the tractor.

Most tractor seats provide for adjustment **backwards or forwards** (horizontal). Some provide for **height or weight adjustment** and some for **back adjustment**. Some seats also provide for a shock-absorbing adjustment. This adjustment can be made more easily while the tractor is in motion.

There are generally **three basic seat adjustments**.

a. Horizontal.
b. Height or Weight.
c. Back.

a. Horizontal

To check the horizontal seat adjustment, proceed as follows:

1. *Check for proper foot control.*
   The horizontal seat adjustment is the most important of the three adjustments you can make while the tractor is standing still. It provides the right distance between you and the steering wheel and between you and the pedals.
   When the seat is properly adjusted, your legs should be straight and your back braced against the back of the tractor seat with either pedal completely depressed. On some tractors, the clutch and brake pedals are not the same distance from the seat; therefore, you have to adjust to the one that requires the greater reach.
   According to some studies, the ideal situation occurs when your leg angles slightly before depressing the pedal (Figure 115) and is straight after your foot has traveled a distance of 7.8 to 10 cm (3¼ to 4 in).
   If the seat is not properly positioned for you, loosen and reposition it until it is in proper position.
   You can usually adjust a seat by **loosening the cap screws, using a lever** on the side or bottom of the seat, or **releasing the seat latch** and adjusting with selector lever (Figure 116).
2. *Check seat position for comfortable steering.*
   Studies show that in the ideal position your elbows are at an angle of 90° as you sit with both hands on the steering wheel (Figure 115). This position enables you to react quickly and to turn the steering wheel through a wide angle. There is also a minimum of fatigue in this position.

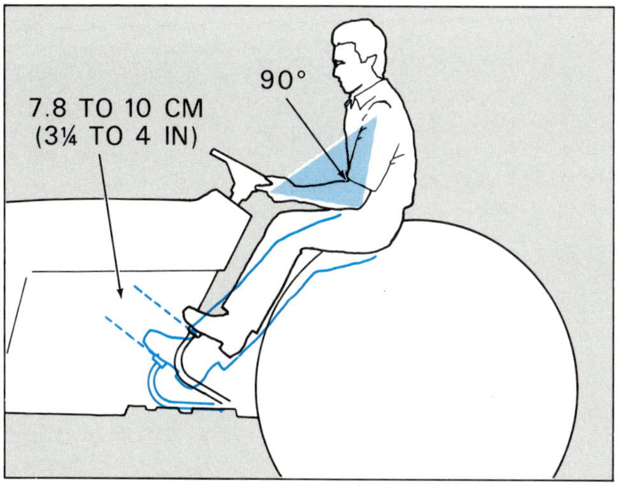

FIGURE 115. Most desirable position of body in relation to steering wheel and pedals.

If the horizontal adjustment does not provide the proper angle for your arms, you may be able to secure it through height adjustment.

b. Height or Weight

To check seat adjustment for height or weight, proceed as follows:

1. *Check for proper foot control.*
   Other than positioning for steering ease (as just discussed), this adjustment is largely a matter of feeling. The position you select should be the one in which you feel most comfortable and have best control of the tractor.

   Adjustments are of two types. With one you remove bolts from the seat supports (Figure 117) and raise or lower the seat to the desired position.

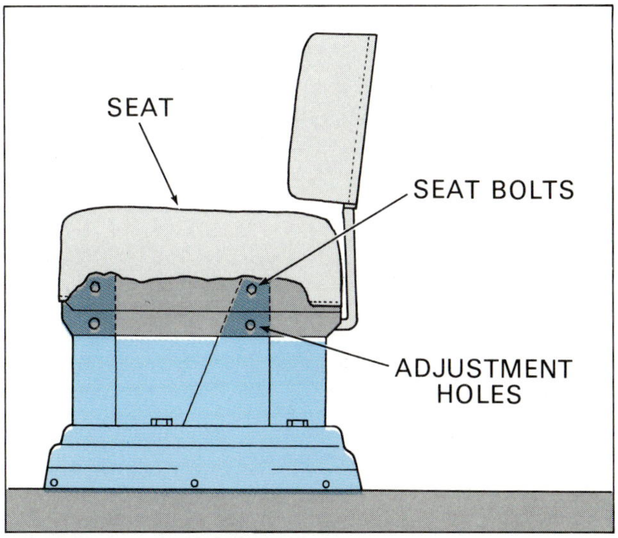

FIGURE 117. Vertical adjustment provided by bolting seat in different positions.

   The second type is a shock-absorber seat—a type that adjusts automatically to your weight. With the same adjustment, a heavier person sits lower than one who weighs less; consequently, persons of varying weight need to reposition the weight adjustment as shown in Figure 118. Other seat adjustments include tilt control, height control and forward and rear control.

2. *Check for comfort.*
3. *Check the seat safety switch if installed.*
   A safety switch has been added to the seat on many lawn and garden tractors built during and after 1973. It is designed to prevent children from operating the tractor. The switch has been set at the factory to engage only when a person weighing 100 pounds or more is seated. The weight required to activate the switch can be increased or decreased. On one type, a bracket is moved up to increase the weight requirement and down to decrease it (Figure 119).

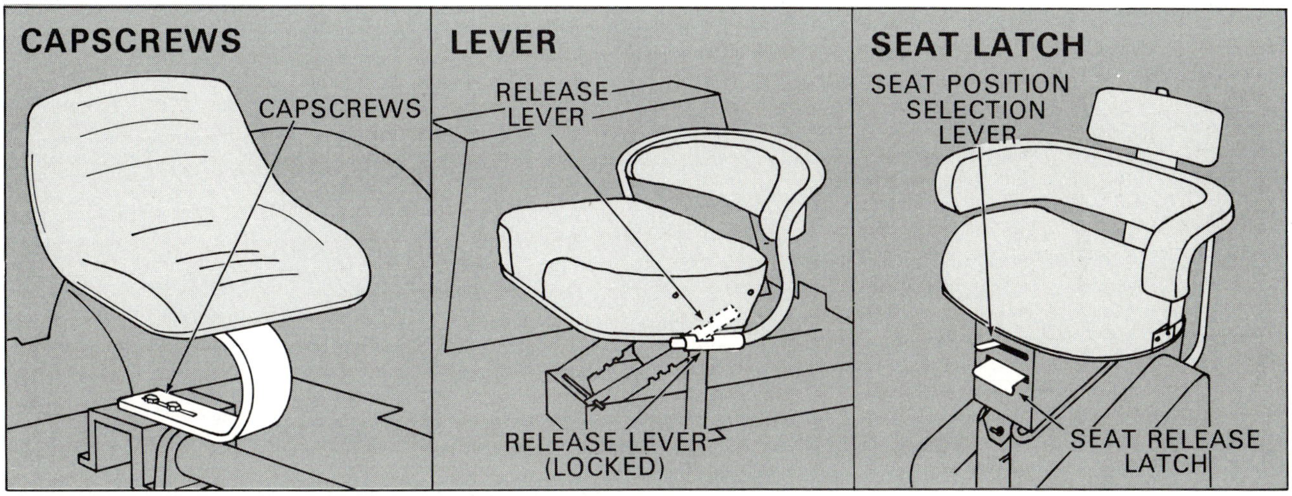

FIGURE 116. Examples of forward and backward seat adjustments.

FIGURE 118. Provision for a weight-adjusting seat.

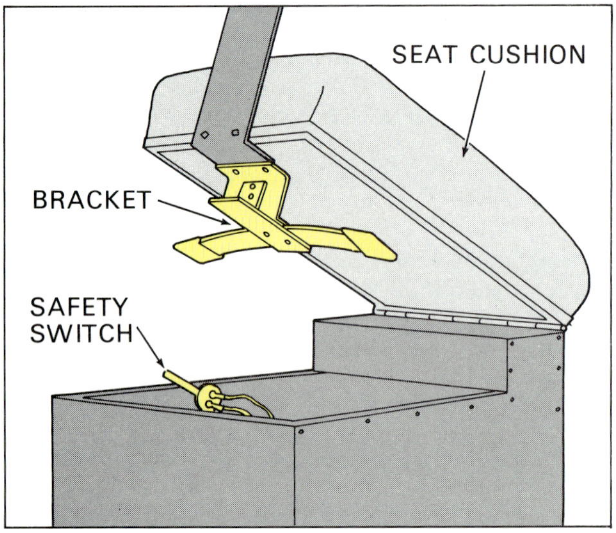

FIGURE 119. Seat equipped with safety switch.

c. Back

To check for proper back adjustment, proceed as follows:

1. *Check for comfort.*
   Not all tractors provide a back adjustment, but where one is provided, adjust the back rest so that it is approximately 24 cm (9½ ins) above the seat). This height provides proper support for your back without your having to hold your spine rigid to secure proper backing (Figure 120).

2. ADJUSTING THE TREAD WIDTH

Tread width must be changed to meet various needs such as row width. Procedures are given as follows:

a. Adjusting Rear-Wheel Spacing.
b. Adjusting Front-Wheel Spacing.

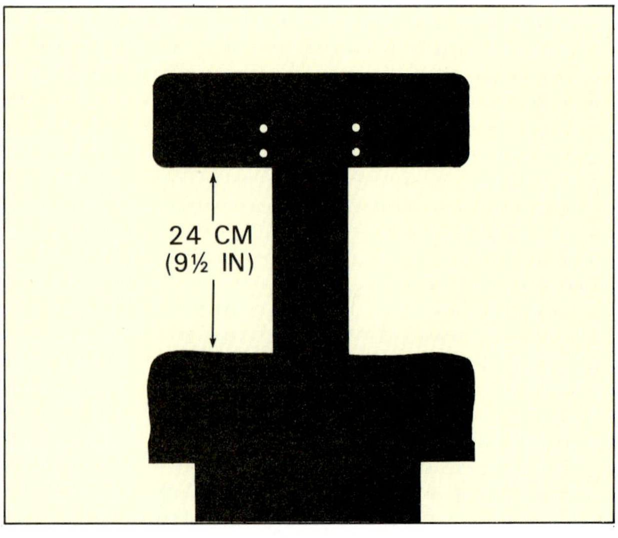

FIGURE 120. Adjust back rest to about 24 cm (9½ ins) above seat to secure proper back support.

a. Adjusting Rear-Wheel Spacing

On most farms, it is necessary to change the rear-wheel spacing several times a year to meet the needs of various farming operations. For example, for plowing, the wheels should be spaced so the right wheels run in the furrow, not on the landside or on the plowed ground. The other wheels must be spaced the same distance from the center of the tractor so that the tractor pulls evenly. For cultivating, the wheels may have to be re-spaced to keep them in the center of the rows. For mounted equipment, the spacing may need changing to allow space for equipment. Close wheel spacing allows more clearance when driving between buildings and through doors and gates. On hills, wide wheel spacing helps avoid overturning.

Adjustable rear wheels can be spaced from widths as narrow as 115 cm (46 in) to more than 250 cm (100 in). Adjustments vary with the make and model of tractor. The adjustments are usually in increments of 10 cm (4 in) that is, each wheel can be moved in or out 5 cm (2 in) at a time or a total of 10 cm (4 in) for both wheels.

Manufacturers provide two means of changing rear wheel spacings:

(1) Manual Adjustment of Rear Wheels.
(2) Power Adjustment of Rear Wheels.

(1) MANUAL ADJUSTMENT OF REAR WHEELS. There are five basic ways of adjusting the spacing of rear wheels by manual methods. They are as follows (Figure 121):

- **Positioning rear-wheel clamps**, either inside or outside of wheel disk.
- **Placing rim** in different position on wheel disk.
- **Changing dish-shaped wheel disks** so wheel will either dish in or dish out.
- **Sliding wheels on axles** to any position desired (if tractor has extended axles).
- **Using offset** or **extended hub**.

Manufacturers use many variations of these methods. The ones shown in Figure 121 are only a sample of those used—those considered as representative of the different methods. If you understand these, you should be able to see readily how other designs work.

Most tractor manufacturers use a combination of two or more of these methods. Your operator's manual shows which combination to use to get the width you desire.

If you have no operator's manual, the following general procedures will work with most tractors:

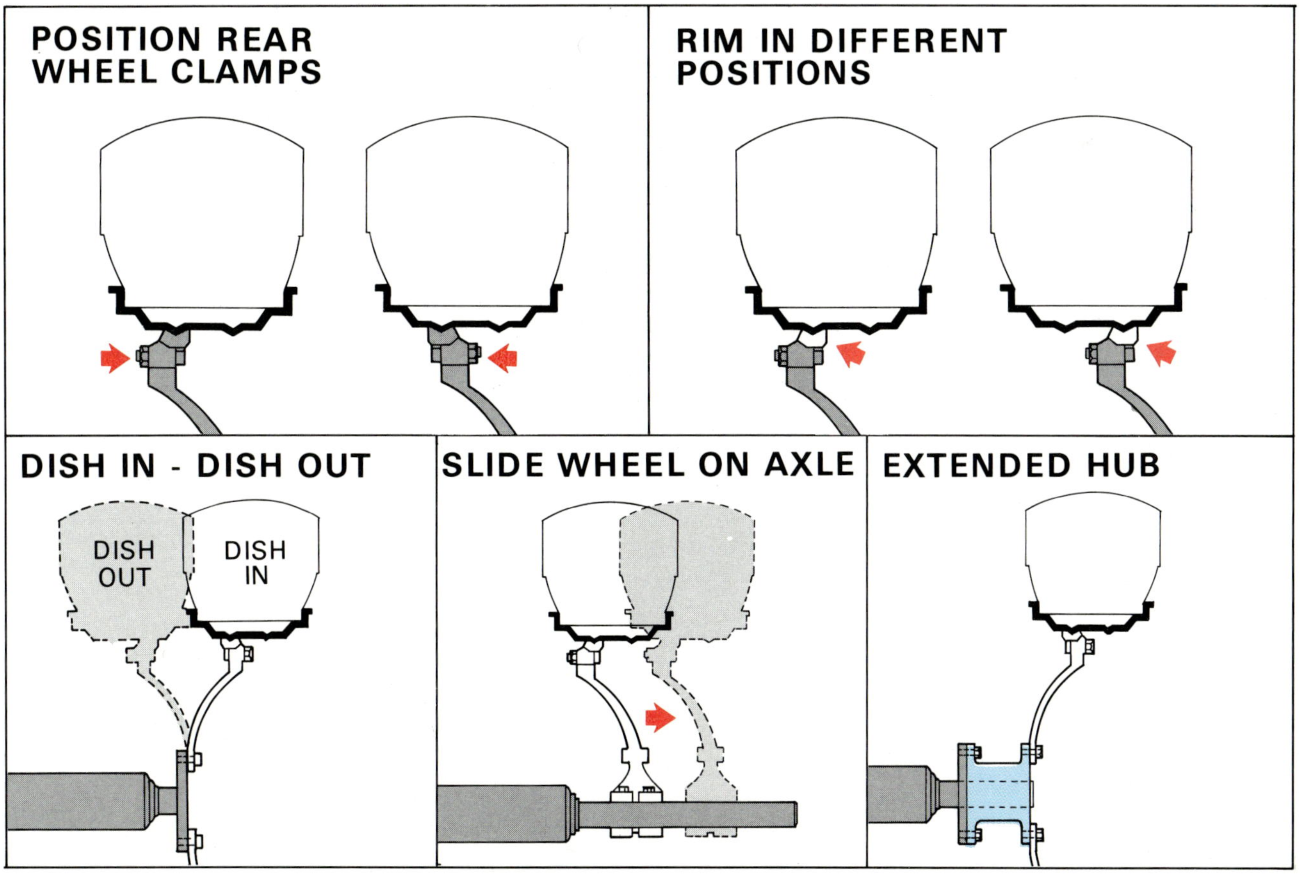

FIGURE 121. Methods commonly used for manually changing rear-wheel spacing.

FIGURE 122. Block front wheels.

1. *Place tractor on a level area.*
   This position helps to keep the tractor from moving after it has been jacked up.
   If you plan to exchange wheels or the tire-and-rim assembly, place tractor near a building, post or tree so that you have something solid to lean the first wheel against while you loosen and remount the second wheel. An overhead hoist is excellent if one is available.
2. *Place blocks in front and back of front wheels to avoid tractor movement (Figure 122).*
3. *Jack up the rear wheels.*
   Place the jack under the axle housing (Figure 123), not under the axle shaft. If placed under the axle shaft, the wheel and axle may be rotated enough to throw the jack out of position and let the tractor fall to the ground.

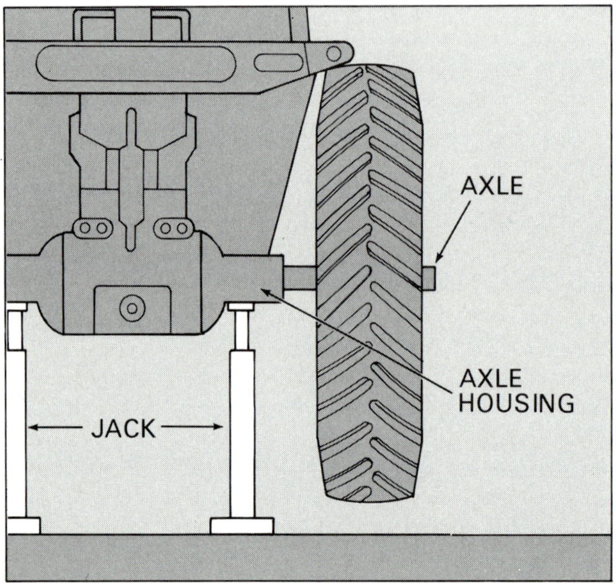

FIGURE 123. Place jack so it lifts rear axle housing.

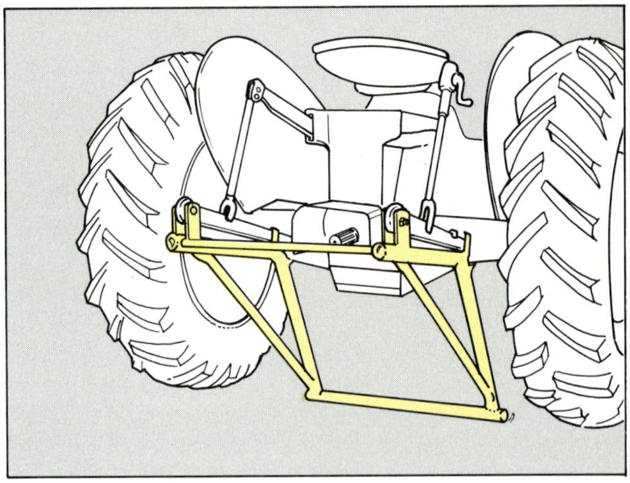

FIGURE 124. Using the hydraulic system to lift tractor rear end.

If the rear wheels are to be changed, jack up both sides. Jacks are available which can be used with your tractor's hydraulic system (Figure 124).

If you have a tractor with a **double-acting hydraulic system**—one that works under pressure whether the hitch is being lifted or lowered—you can place blocks under the hitch points of the lower links and lower the hitch to lift the rear wheels.

4. *Examine wheel assemblies to determine which of the following methods, or combinations of methods, you will need to use for changing wheel spacing.*
   (a) Changing Wheel-Clamp Positions.
   (b) Changing Rim Position on Wheels.
   (c) Exchanging Rear Wheels.
   (d) Sliding Wheels on Extended Axle.
   (e) Using the Offset Hub.

(a) Changing Wheel-Clamp Positions. Proceed as follows:

(1) *Lock wheel brakes (Figure 125).*

FIGURE 125. Always lock wheel brakes.

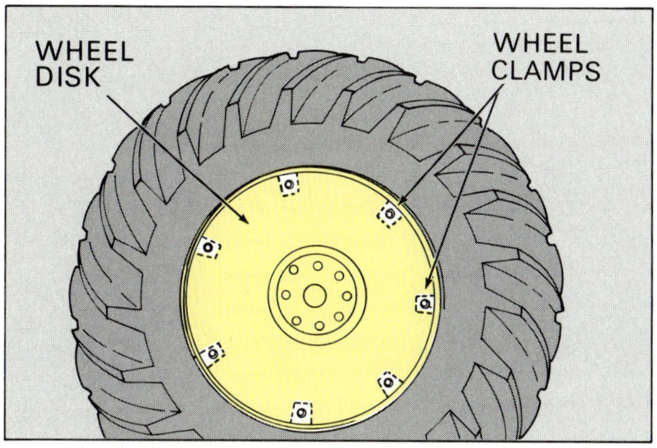

FIGURE 126. Remove all wheel clamps.

(2) *Remove all wheel clamps and let tire rest upright against wheel disk (Figure 126).*

(3) *Reverse and reinstall wheel clamps in upper position of wheel disk first (Figure 127).*

(4) *Place tire-and-rim assembly on newly-positioned wheel clamps (Figure 128).*

(5) *Install remaining wheel clamps and tighten evenly.*

(b) Changing Rim Position on Wheels. Proceed as follows:

(1) *Turn rear wheels until drive clamps are on upper side (Figure 129).*
Tractors equipped for changing rim positions usually have two special clamps that guide the rim as it is being moved from one position to the other.
Most rims have a bead on each side of each drive clamp that extends crosswise. The beads act as a guide to keep the rim in position on the drive clamps.

(2) *Remove mounting clamps.*
Do not remove drive clamps.

(3) *Use pry bar to move rim to desired position on wheel (Figure 130).*

(4) *Replace mounting clamps and tighten securely.*

FIGURE 127. Wheel clamps hold rim to disk.

FIGURE 128. Replace tire-and-rim assembly.

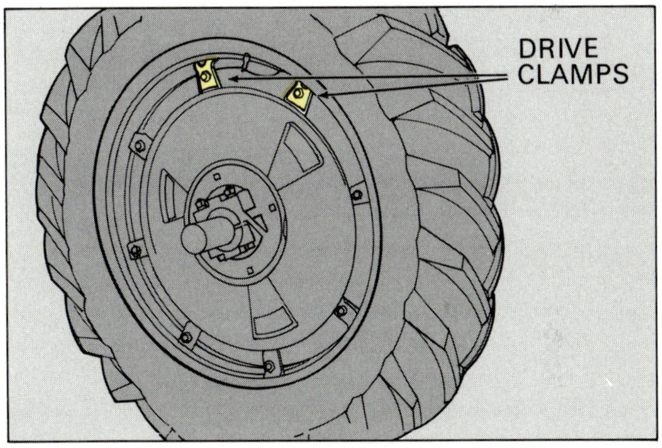

FIGURE 129. Drive clamps should be on upper side of wheel.

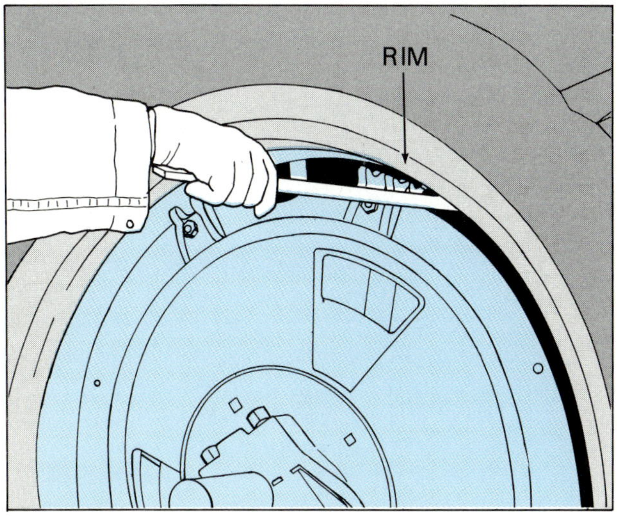

FIGURE 130. Use pry bar to position rim.

FIGURE 131. Remove wheel bolts.

(c) Exchanging Rear Wheels. Proceed as follows:

1. *Remove wheel bolts on one wheel (Figure 131).*
2. *Remove wheel and leave in upright position.*

   Lean wheel against a building, post or tree or hold with an overhead hoist (Figure 132).

   Tie top of wheel to the support to keep it from rolling or falling if you, or a child or animal should accidentally hit it.

   Don't lean the wheel against the back or side of the tractor. There is too much risk of pushing the tractor off the jacks.

FIGURE 132. Great care should be taken when handling wheels.

Don't try to lay the wheel down. Most wheels are very heavy, and you may get hurt. Also, you will need help to get the wheel raised to an upright position again. A wheel often weighs 225 to 450 kg (500 to 1,000 lbs).

(3) *Remove second wheel and place on hub or axle of first wheel.*

Be sure to exchange wheels. The tire lug bars are chevron-shaped. The chevrons on top of the tire should point in the direction of forward travel in order to get satisfactory traction and self-cleaning action of the lugs. Also, look for a "direction-of-rotation" arrow molded on the side of the tire.

There is one exception to the above statement. Offset tractors—ones with the engine mounted to the left side of the tractor chassis—are equipped with a cast-iron wheel on the right side to balance the engine weight. With these, remove disk and turn so that it will dish in the opposite direction, mount on axle and replace tire-and-rim assembly.

(4) *Tighten wheel bolts on both wheels.*

(d) Sliding Wheels on Extended Axle. Proceed as follows:

(1) *Rotate wheel until keyway is on upper side of shaft (Figure 133).*

One make of tractor uses a rack and pinion. In that case, rotate wheel until rack is in upper position.

(2) *Loosen hub-clamp bolts.*

(3) *Slide entire wheel assembly to desired position on axle.*

If your tractor has a rack-and-pinion adjustment, loosen three cap screws and turn jack screws clockwise until outer groove in each screw is flush with outer surface of wheel hub. The wheel is moved on axle by using flywheel wrench to work the rack-and-pinion assembly (Figure 134).

(4) *Tighten hub bolts securely.*

With rack-and-pinion design, loosen jack screws and tighten cap screws in order to wedge wheel against axle.

(e) Using the Offset Hub. Proceed as follows:

(1) *Use steps 1 and 2 under (d).*

(2) *Follow steps 2 and 3 under (c).*

(3) *Follow steps 3 and 4 under (d).*

(4) *Tighten bolts again after a few hours of operation.*

This procedure is important. Every year tractor operators ruin wheels, hubs, rims and axles because they don't tighten

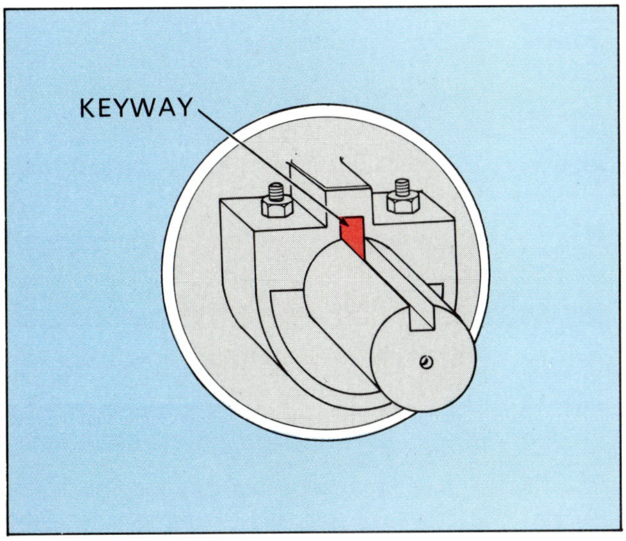

FIGURE 133. Make sure keyway is on upper side of shaft.

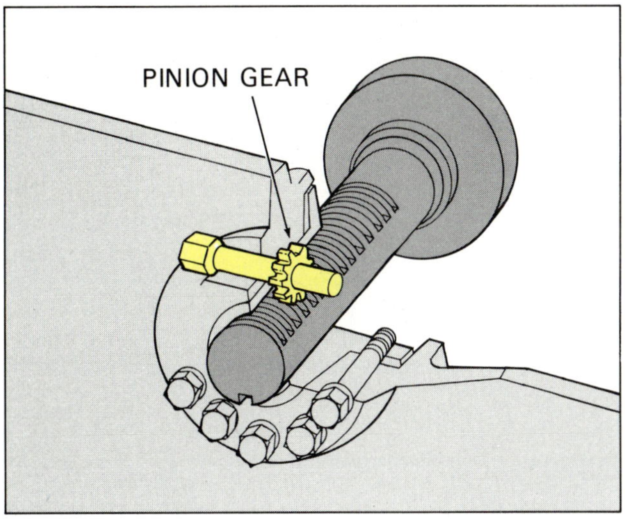

FIGURE 134. Slide wheel assembly to desired position.

wheel bolts and nuts sufficiently. Sometimes they aren't tightened enough when the wheel spacing is first changed even though they appear tight. Other times, looseness shows up after the tractor has been in operation for several hours.

(5) *Measure distance from center of tractor to center of tire tread (Figure 135).*

FIGURE 135. Measure distance from center of tractor to center of each tire tread to make sure they are equally spaced.

Some mounted implements may require one wheel to be set closer to the tractor than the other, but for general use, each tire should be an equal distance from the center of the tractor.

(2) POWER ADJUSTMENT OF REAR WHEELS. Even though a tractor has provision for power adjusting its wheel spacings, it usually has provision for manual adjustments as well. For example, suppose a tractor is equipped with an extended axle and power-adjusting wheels. Assume the wheels are set so that you can power adjust them from a width of 132 to 193 cm (52 to 76 in). If you need 203 cm (80 in), you will have to move each wheel out 5 cm (2 in) on its axle. Your new range of power adjustment will then be between 142 and 203 cm (56 and 80 in).

Once the wheels are set for your minimum and maximum spacing needs, you will usually have enough selection of wheel spacings in the power-adjustment range to meet all conditions.

There are two types of power-adjusted rear wheels. Although the two types look different, they work on the same general principle. Both depend on some provision for the wheel disk to rotate inside the rim. They are described as follows:

(a) Spiral Guide-Rail Type.

(b) Slotted Channel-Bar Type.

(a) Spiral Guide-Rail Type. To get a better understanding of the power operating principle, note the spiral guide-rail type in Figure 136. The guide rail (sometimes called a "shift" rail) is welded to the rim. There may be 4 to 6 guide rails on a rim.

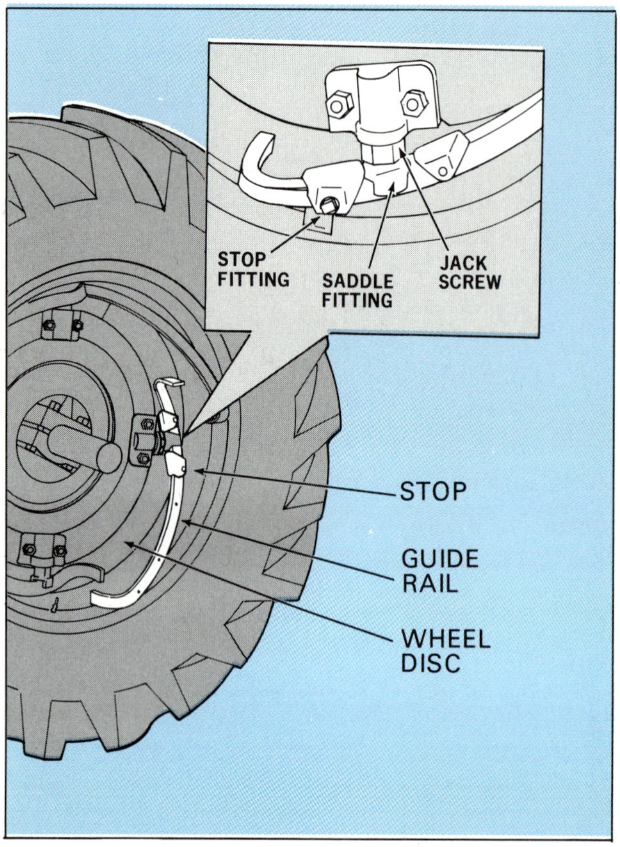

FIGURE 136. A spiral guide-rail type of power adjusted wheel. A jack screw clamps the wheel disk to the guide rail and rim.

Between each rail and the wheel disk is a **jack screw**. Each jack screw is equipped with a **saddle fitting** that rides the **guide rail**. When the jack screws are loosened, the wheel disk is free to slide on the guide rails. Since the guide rails are spiral-shaped, rotating the wheel disk forces the tire-and-rim assembly in or out, depending on which direction the wheel disk moves on the guide rail.

Instead of a jack screw, some tractors are equipped with **eccentrics** (Figure 137) which can be loosened or tightened against the rail to perform the same function as a jack screw.

On each wheel there is one guide rail with holes in it. This is the one equipped with two **stop fittings**. The stop fittings are used to control the wheel-space setting as the following procedures show.

Your operator's manual gives detailed procedures for power adjustment. In general, the following procedures apply (Figure 138).

NOTE! Adjust one wheel at a time.

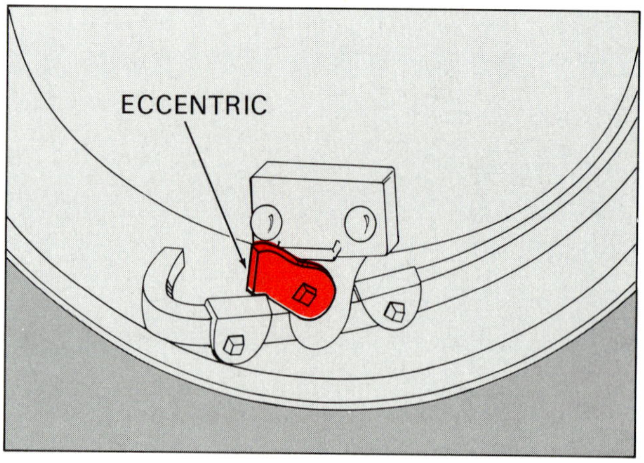

FIGURE 137. Eccentric used on some tractors in place of jack screw.

1. *Determine which way saddle fitting must move on rail to secure desired adjustment.*
2. *See if there is enough remaining space on guide rail to provide the adjustment desired.*

   Each wheel has 25 to 30 cm (10 to 12 in) of lateral adjustment from one extreme position to the other. Locate the guide rail with adjustment holes and count the number of spaces left in the direction you want the wheel to move. Most wheels provide for 5 cm (2 in) of lateral movement, but some provide 6¼ cm (2½ in).

   For **maximum adjustment**, some tractors use the space between the last hole and the end of the rail. If there isn't sufficient adjustment, you will need to use one of the manual adjustments provided.

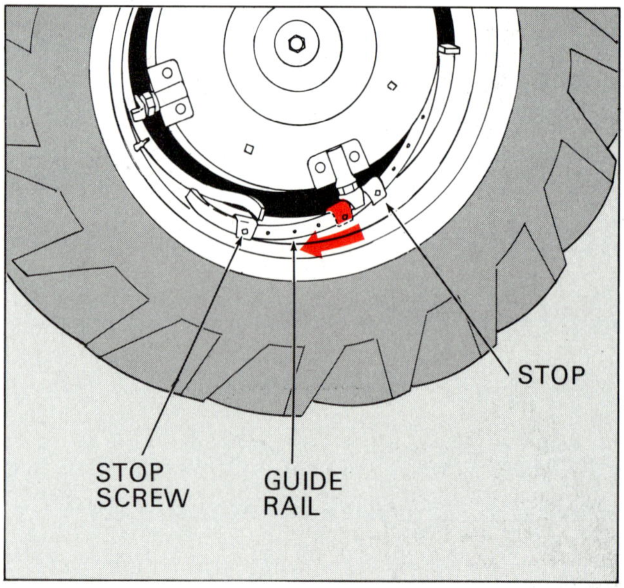

FIGURE 138. Stop is moved to position desired for new tread width.

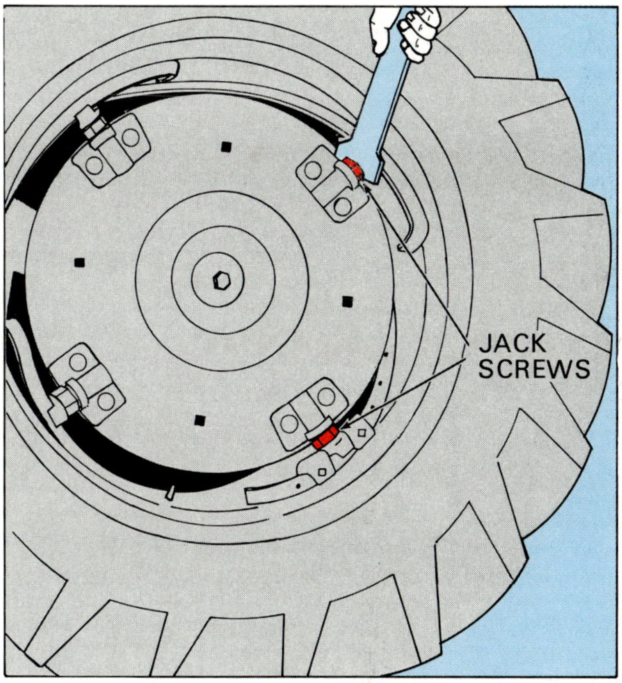

FIGURE 139. Some tractors are equipped with four jack screws (or eccentrics) while others have six.

3. *Remove stop on rail to permit movement in desired direction.*
4. *Replace stop in rail hole that provides the desired wheel spacing and tighten set screw.*
   If tread width is being narrowed as much as possible, check to make sure that rim and tire will not interfere with the fender or belt pulley.
5. *Loosen nuts on the two (or three) upper jack screws an equal number of turns.*
   If eccentrics are used in place of jack screws, loosen all of them.

   Some tractors are equipped with four jack screws while others have six (Figure 139). If your tractor has four, loosen two upper ones. If it has six, loosen three upper ones. Then rotate wheels ½ turn by driving tractor and loosen remaining jack-screws. The principal advantage of this suggestion is to avoid working against the weight of the tractor. Loosen all the screws until the wheel disk is free to rotate.

   Loosening screws an equal number of turns and retightening in the same way keeps the wheel centered on the axle.
6. *Start tractor and shift into gear.*
   Use first gear or reverse, depending on which way you want wheel disk to rotate.
7. *Apply brake on side not being adjusted and slowly engage clutch.*
   Engage clutch until jack screw or eccentric contacts the stop you reset on the rail (Figure 140). This step may require about ¼ turn of disk.

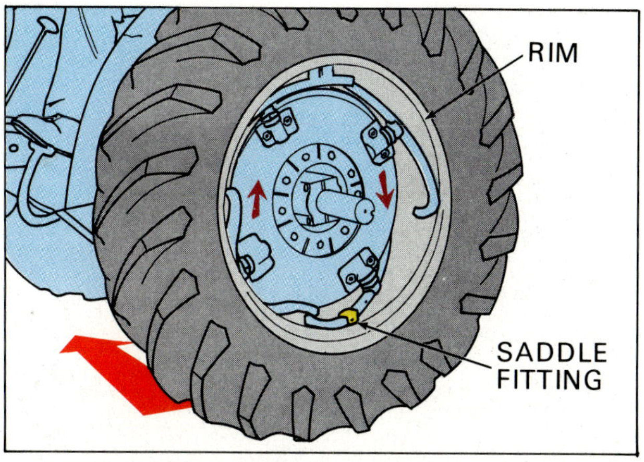

FIGURE 140. With tractor shifted into reverse gear, wheel disk turns backwards, saddle fittings move towards outside of rim pulling rim-and-wheel assembly towards tractor.

8. *Move second stop against jack screw or eccentric and tighten securely (Figure 141).*
9. *Tighten the two (or three) jack screws which were loosened, or if equipped with eccentrics, tighten all of them.*
   To tighten the equipment properly, most manuals recommend a pull of approximately 64 to 66 kg (150 to 160 lbs) on end of wrench provided for jack screw adjustment.
10. *Proceed with adjustment on second wheel in same manner.*

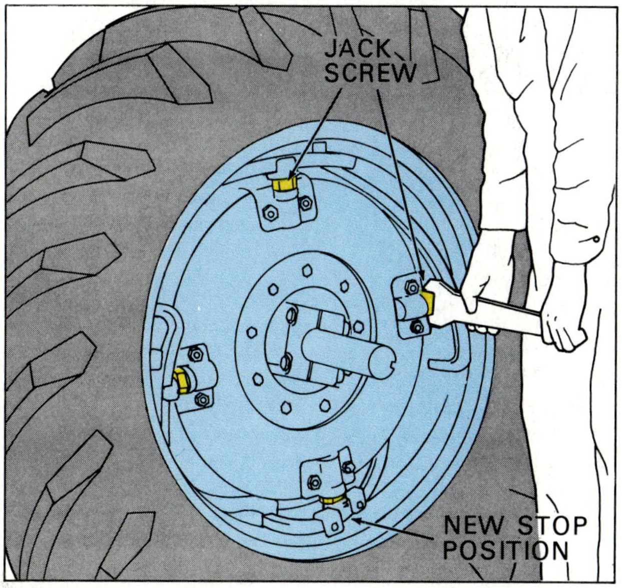

FIGURE 141. Move second stop to new position and tighten. Tighten loosened jack screws or eccentrics.

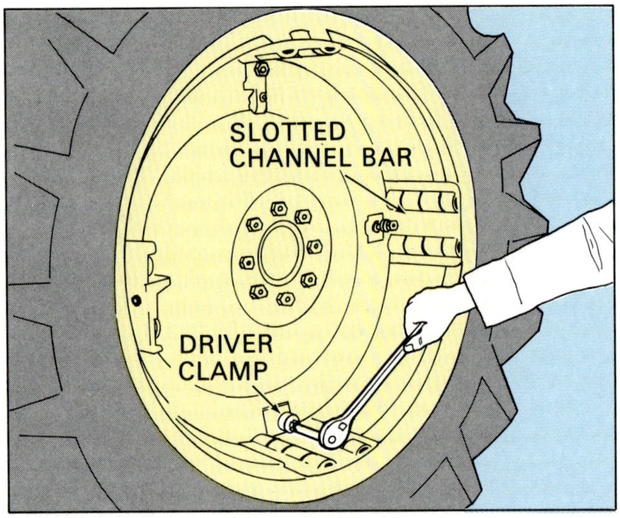

FIGURE 142. With the slotted channel-bar type of adjustment, the wheel disk threads through the slotted channel bar in same manner as bolt threads into nut.

the rim. As power is applied to the disk, it rotates and threads through slots in the channel bar in the same manner a bolt threads into a nut (Figure 142). The wheel disk makes a complete turn for each 5-cm (2-in) adjustment. General procedures for using this type of power adjustment are as follows:

1. *Rotate tractor wheel so spacer clamp is at top.*
   The spacer clamp is different in appearance from the other clamps (Figure 143). It has a small bolt in the arm that projects from it. It is designed for two uses: (a) it serves as a driver clamp to hold wheel disk and rim together except when wheel spacing is changed; then (b) it is used as a stop.
2. *Loosen large nut that holds spacer clamp on wheel disk.*

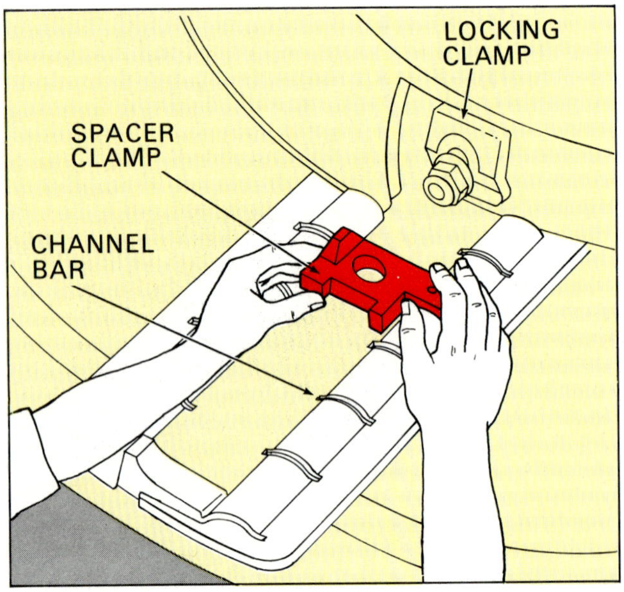

FIGURE 143. Spacer clamp.

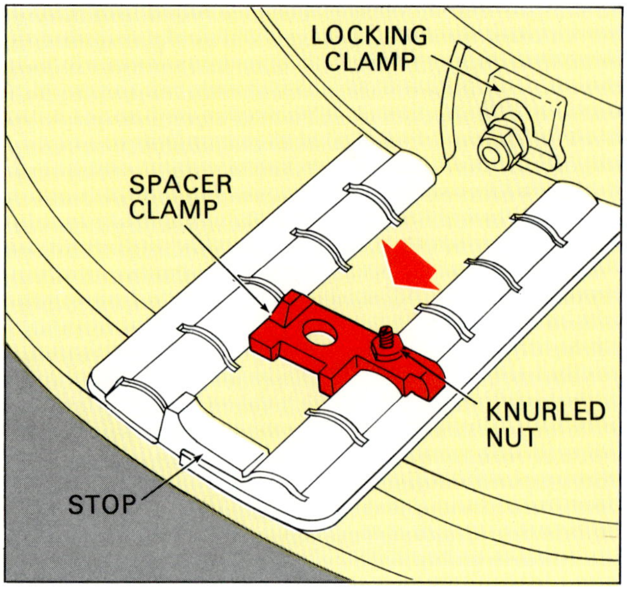

FIGURE 144. Spacer clamp is placed in channel bar at point where you desire wheel disk to stop.

3. *Loosen small knurled nut in spacer-clamp arm. Turn bolt so special head will pass through opening in wheel disk.*
4. *Loosen nuts on remaining lock clamps, slide away from rim and retighten to wheel disk.*
   With all lock clamps retracted, the disk is free to turn independently on the rim.
5. *Place spacer clamp in channel bar at desired wheel spacing (Figure 144).*
   Insert oval-headed bolt (in spacer-clamp arm) into the channel-bar slot and turn to secure the bolt head in the channel. Tighten knurled nut securely.
6. *Start engine and shift into gear.*
   Use first gear or reverse, depending on which way you want wheel disk to rotate.
7. *Engage clutch and, at same time, brake opposite wheel lightly.*
   Light braking of opposite wheel is to let tractor move ahead slightly.
8. *Disengage clutch when wheel disk strikes spacer clamp (Figure 145).*
9. *Rotate tractor wheel until lug for spacer clamp is at top.*
10. *Remove spacer clamp from channel bar and replace on lug for use as a fourth wheel clamp.*
11. *Loosen nuts on remaining lock clamps, position in rim channel bar and tighten against rim.*
12. *Use the same procedures in changing tread width on second rear wheel.*
    Be sure to check the jack screws, eccentrics or clamps after a few hours' operation. They sometimes become loose and can cause damage to the wheel-adjustment fittings.

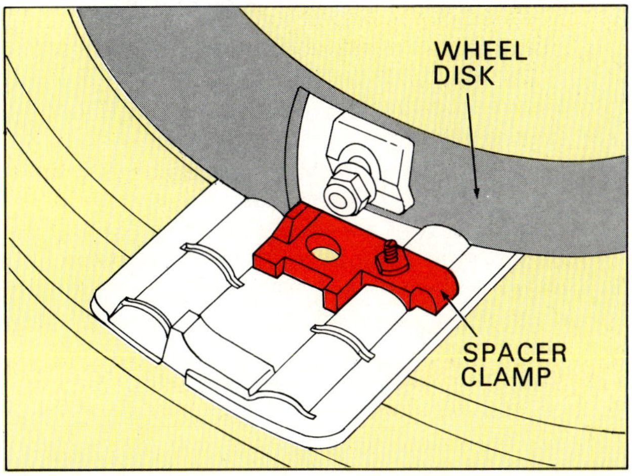

FIGURE 145. Wheel disk is rotated until it contacts spacer clamp.

b. Adjusting Front-Wheel Spacing

As you look at different tractors, you will note there are three different types of front wheel arrangements (Figure 146):

(1) Single front wheel.
(2) Dual front wheels.
(3) Adjustable front axle.

Of the three, only the adjustable-axle type needs to be matched with the spacing of the rear wheels.

(1) The **single-wheel type** has no spacing adjustments.

(2) The **dual-wheel type** is normally run with a narrow setting. However, the wide setting is an advantage under two conditions: in mud there is less tendency for balling up between the wheels, and with listed crops it is easier to keep the front wheels on the ridge. Wider spacing is provided either by reversing the disk-shaped wheels or by changing the rim position from one side of the wheel disk to the other (as provided with manually-spaced rear wheels).

(3) If you are using a tractor with an **adjustable front axle**, there are several front-wheel spacings you can use to match the spacing of the rear wheels.

The spacing between the front wheels can be changed in units of 10 cm (4 in). Some provide 12.5-cm (5-in) spacings to match the rear wheel width. If only a small amount of width increase is needed, the front wheels can either be reversed (of the disk type) or mounted on the opposite side of the wheel disk, to supply from 10 to 20 cm (4 to 8 in) of additional wheel spacing depending on design. This change has no effect of steering as with dual wheels.

Sometimes it is desirable to set one wheel out farther than the other. Two examples are as follows: (1) when you wish to use one wheel as a marking or guide

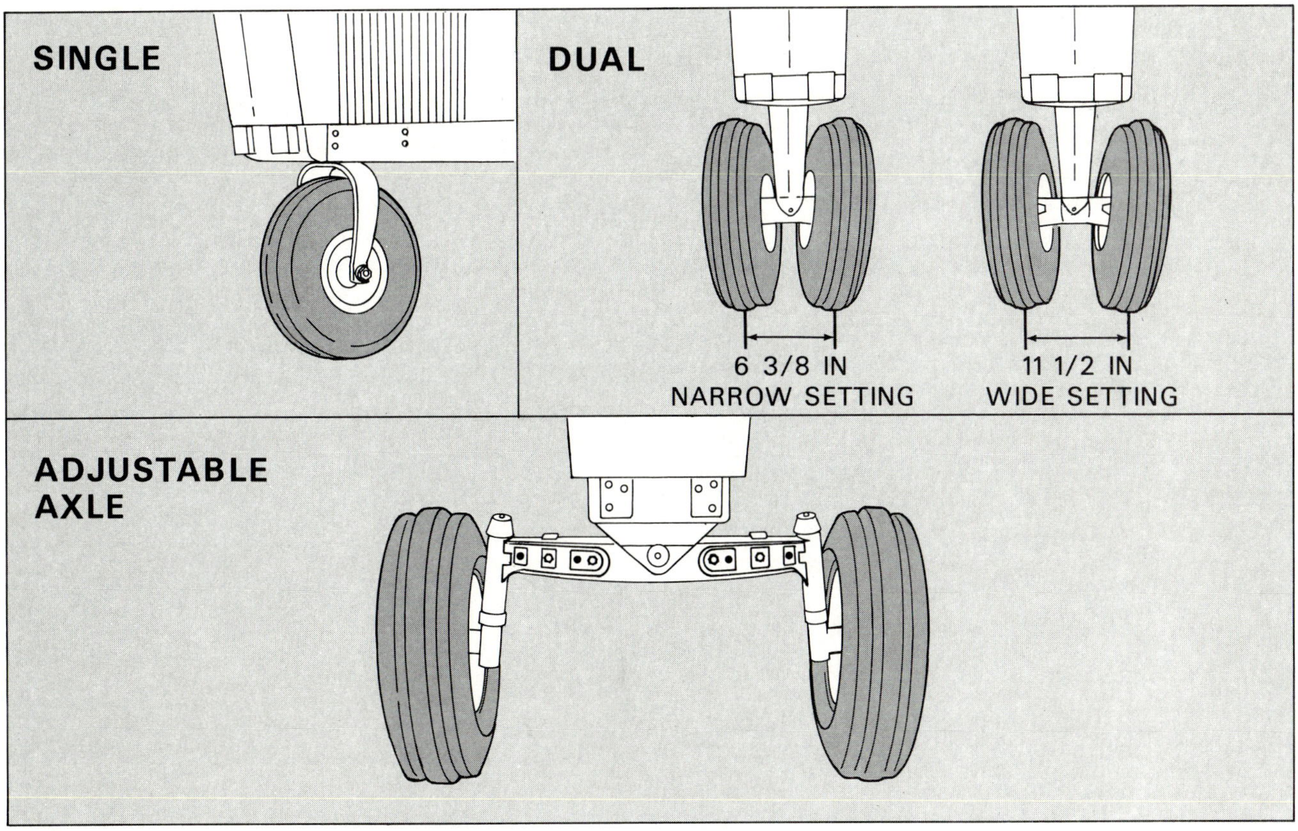

FIGURE 146. Types of front wheel arrangements.

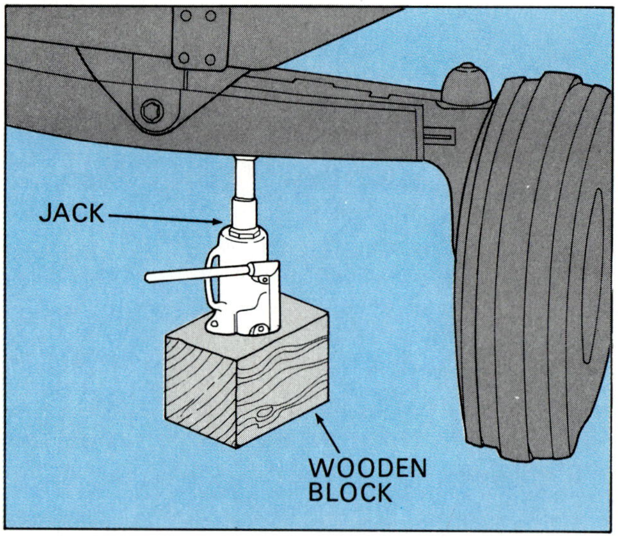

FIGURE 147. Always use a good jack with a wide base when jacking up the front end of a tractor.

wheel during seeding or planting, and (2) if you have mounted equipment, such as a corn picker, it may require more space on one side than the other.

As you look at the adjustable front-axles of different tractors, they may appear quite different in design. In general, though the procedures for extending or shortening axles are similar. Your operator's manual is recommended for giving exact procedures for your particular tractor. In general, the following apply to all farm tractors.

1. *Be sure tractor is on level ground.*
   This precaution protects you from the possibility of the tractor's rolling off the jack. It also makes the job of adjusting easier.
2. *Lock brakes on rear wheels or place blocks in front and back of rear tires.*
3. *Jack up front end of tractor.*
   Don't try to get by with a makeshift jack. Use a good jack with a wide base (Figure 147).
4. *Loosen both extension axles.*
   The extension axle may be fastened to the center of axle with bolts or clamps, or the two axles may be telescoped and bolted or clamped in position (Figure 148).
5. *Loosen steering linkage (tie rods or drag links) that attach to extension axle.*
   Select the front axle design from Figure 149 that is closest to that of your tractor. Loosening a tie rod or drag link is usually a matter of backing off a set screw and lock nut or removing a clamp bolt.
6. *Slide axle to desired position and fasten.*
   Usually the inside face of the front tire is aligned with the inside face of the rear tire.

   However, you can align the center of the front tires with the center of the rear tires if you wish. There is no advantage of one over the other except when plowing. If you align the inside faces, the tractor will guide itself along a furrow wall.

   Hole positions in the axles are at 5-cm (2-in) intervals on most tractors, so it is easy to determine the distance you are moving each extension axle. (Tractors with 6¼-cm (2½-in) intervals on rear wheels have same intervals on front axle.)

   Fastening is a matter of replacing bolts or tightening clamps that you removed or loosened in step 4. If the extension axle is being bolted to the center axle (Figure 148), leave at least one hole between the two bolts to make the axle more rigid. If your tractor axle assembly is similar to Figure 149, it may have a provision for resetting the **radius rods** to a wide or narrow position to match a wide or narrow wheel spacing.
7. *Adjust tie rods on drag link(s) and tighten.*
   If your tractor assembly has **front or rear tie rods**, you will find the telescoping member has punch marks, holes or ring grooves to mark 5- or 6¼-cm (2- or 2½-in) adjustments. Move tie rod or drag link the same number of notches you moved the extension axle and tighten.

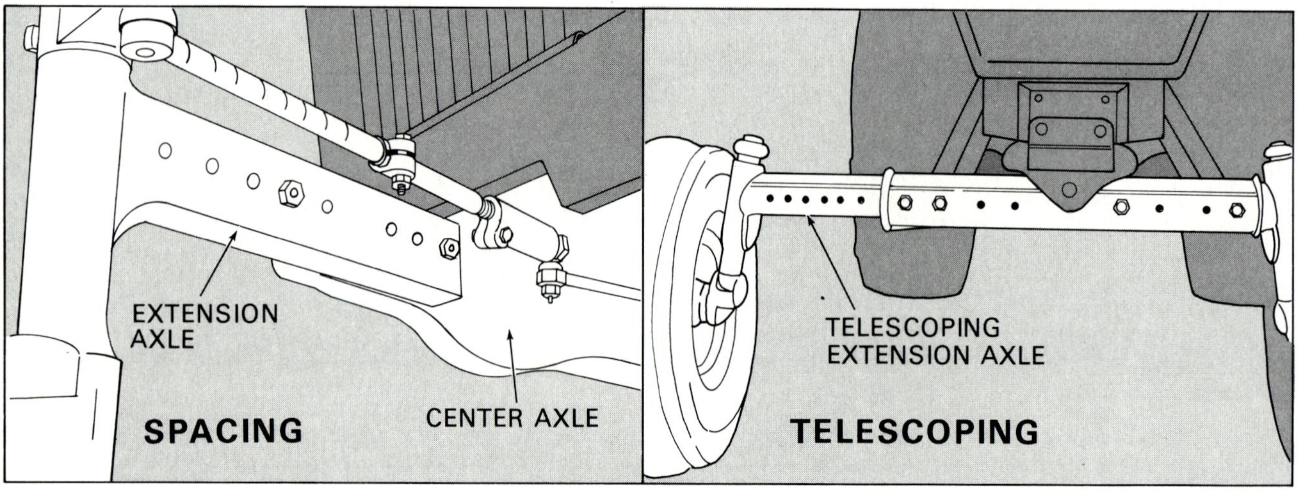

FIGURE 148. Front axle extensions may be adjusted by spacing or telescoping.

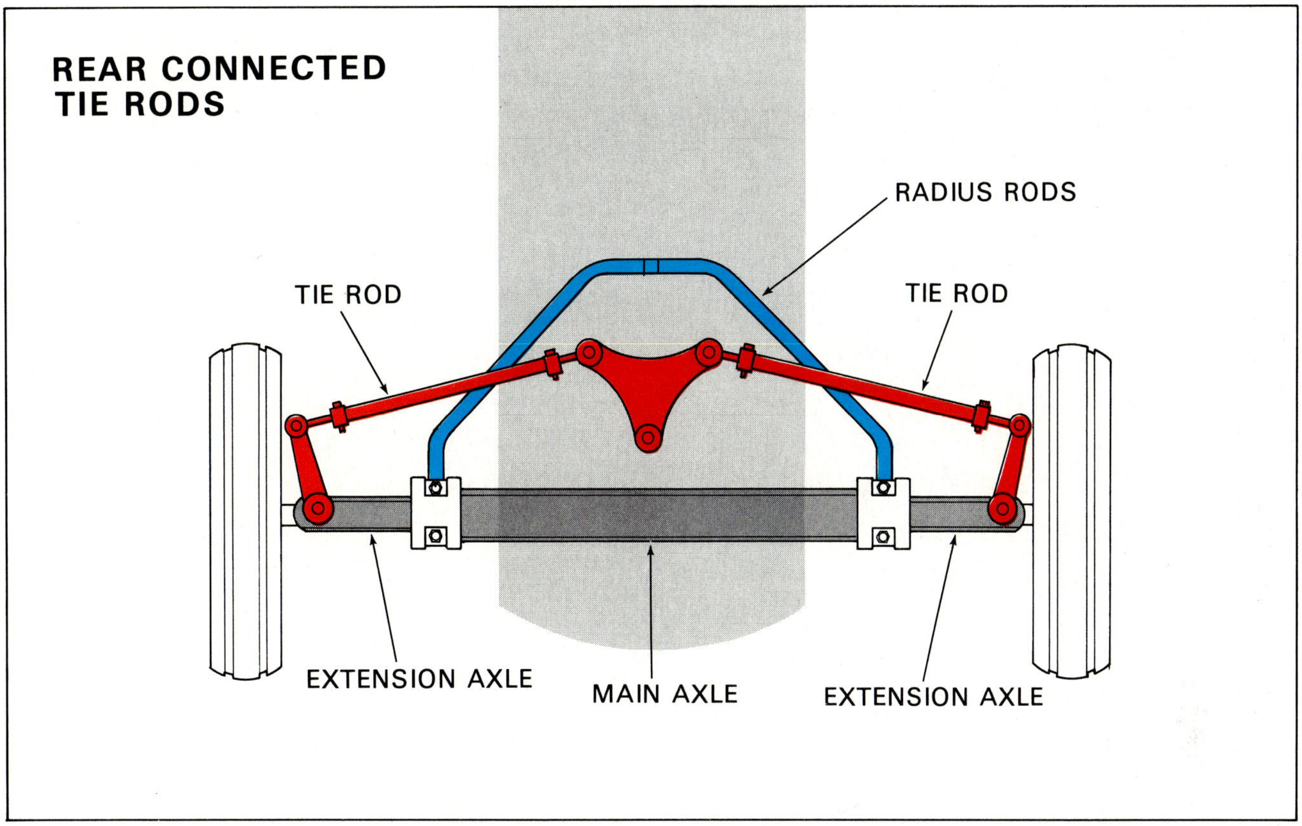

FIGURE 149. An adjustable type front-axle and steering assembly.

If your tractor has an angular axle, check to see if the front end of the radius rod is in line with the front of the drag link. If in line, no adjustment is necessary. If out of line, proceed as follows:

(1) *Center the steering wheel so that the front wheels point straight ahead.*

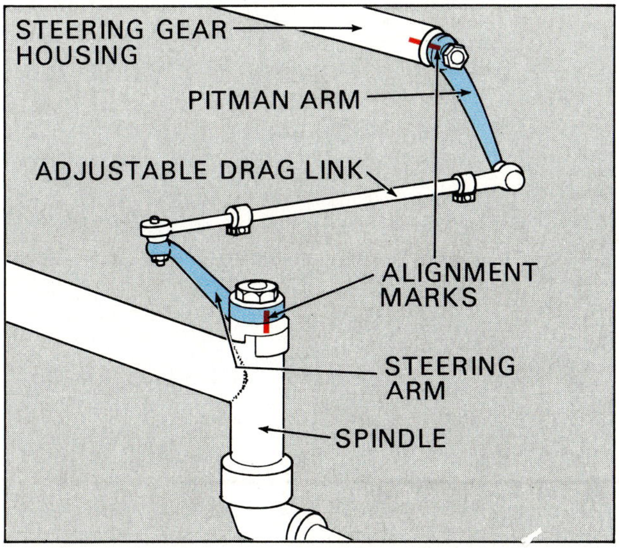

FIGURE 150. Alignment marks commonly used with angular-axle assemblies to secure proper toe-in after changes have been made in front wheel spacing.

You may find a marking on the steering-gear housing and on the pitman arm (Figure 150). If so, align the marks for proper wheel position.

(2) Turn adjustable drag link (turnbuckle) to secure proper toe-in adjustment and tighten. Note align marks on steering arm and spindle (Figure 150). Adjust drag links until these marks are in line.

(3) *Repeat procedures with drag links on opposite side of tractor.*
For further information on adjusting tie rods consult your operator's manual.

3. ADDING WEIGHT

It is a common practice with farm and industrial tractors to use additional weight on the rear wheels and often on the front wheels. The need for extra weight increases as tractors become more powerful in relation to their weight. No matter how much power your tractor has, its power can't be used to advantage when the rear wheels spin.

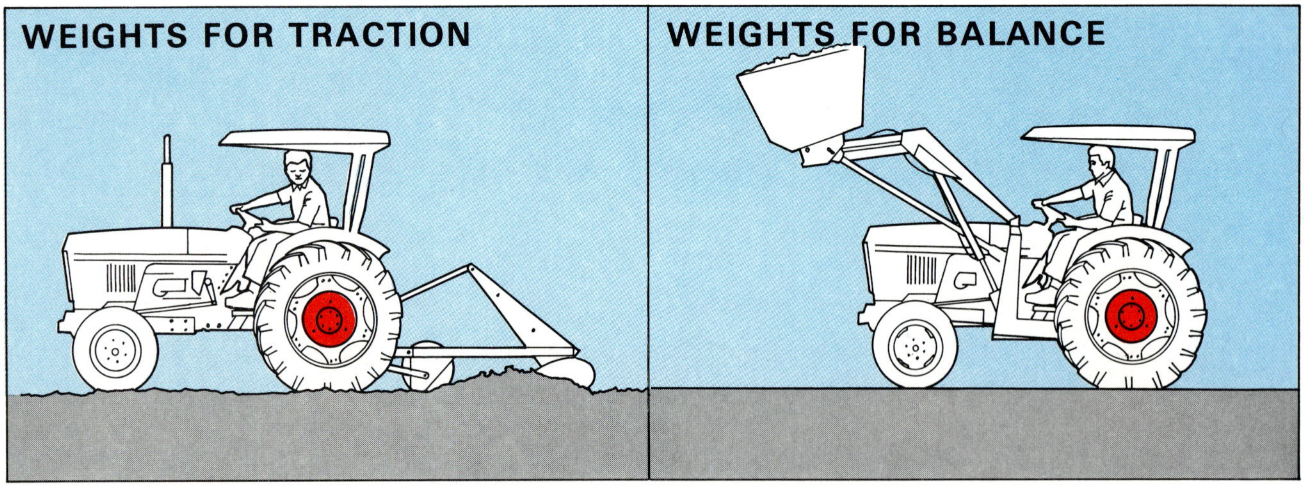

FIGURE 151. Weight is needed on rear wheels to give added traction and to balance heavy front loads.

The reasons for adding weight to **rear wheels** are as follows (Figure 151):

- **To give added traction for pulling**.
- **To maintain traction with a heavy load on the front**.

The reasons for adding weight to the **front wheels** are as follows (Figure 152):

- **To balance the lifting action of the front wheels caused by rear-mounted equipment.**
- **To help keep front wheels of tractor from rising when driving up a slope**.

The drawbar pull of your rubber-tired tractor is usually considered about equal to half the weight carried on the rear wheels. This includes not only tractor weight but also weight from mounted equipment, the driver, etc. Studies conducted by USDA show this is approximately true when operating a tractor on dry clay or sandy loam soils. But with a concrete surface, the drawbar pull increases to about 30 to 32 kg (65 to 70 lbs) per 45 kg (100 lbs) added to the rear wheels. On dry sand or green alfalfa, drawbar pull drops to about 16 to 18 kg (35 to 40 lbs).

With a small tire, adding liquid usually increases the weight 91 kg (200 lbs) or more. Large tires may increase in weight by 250 to 272 kg (550 to 600 lbs).

Under normal heavy-load operating conditions, there is some slippage of the rear wheels—probably about 15 percent, which is considered the maximum allowable under most surface conditions. But as the load increases, wheel slippage can increase (beyond 15 percent) enough to greatly increase tire wear, causing

FIGURE 152. Weight on front wheels is used to offset lifting effect from mounted equipment on rear of tractor and to keep front wheels on ground when pulling up a slope.

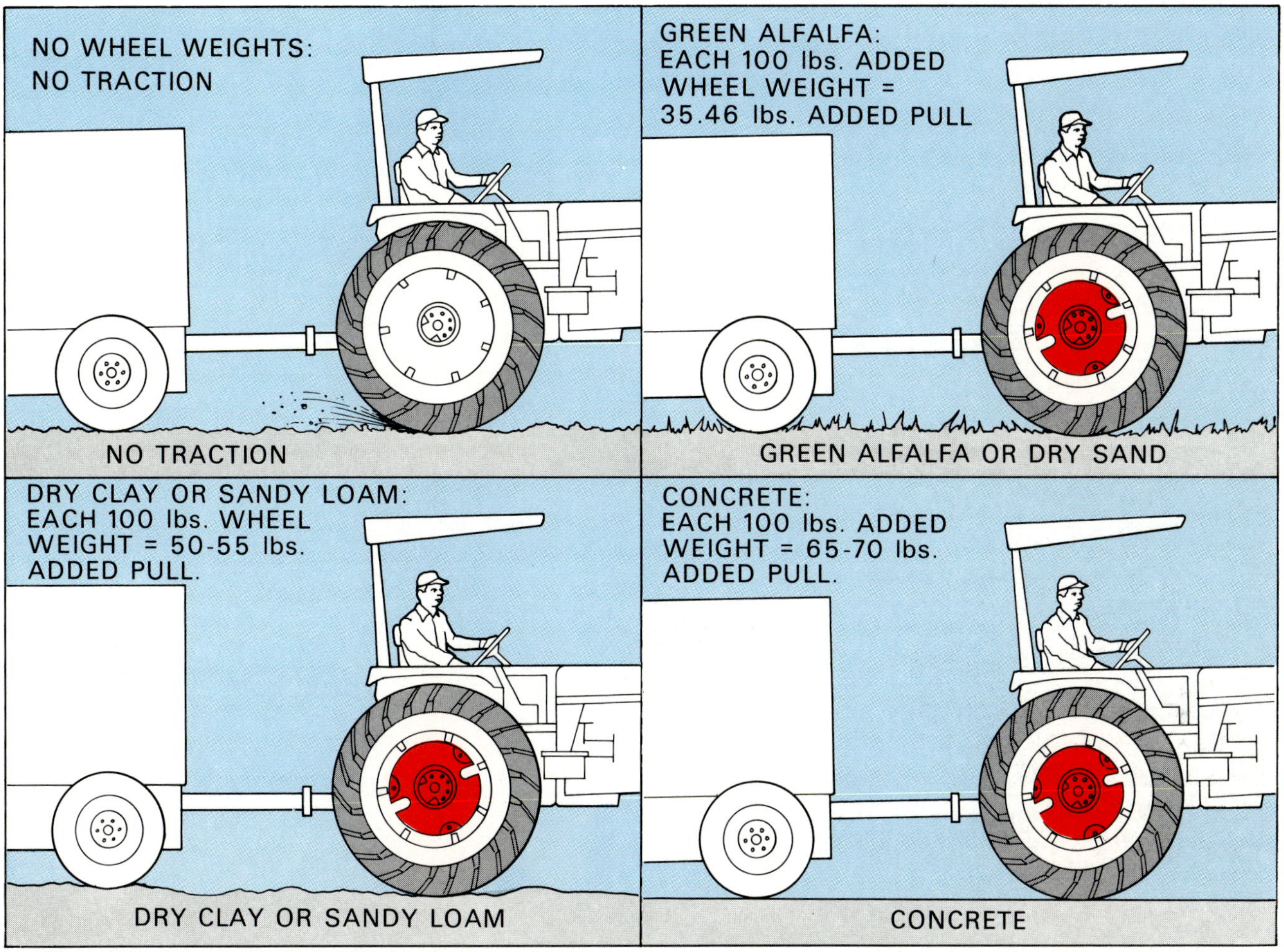

FIGURE 153. A tractor may have ample power but not be able to pull a load because of insufficient traction. Adding weight to rear wheels helps overcome this problem. For each 45 kg (110 lbs) added, drawbar pull is increased about 156 to 178 N (35 to 40 lbs) on dry sand or green alfalfa, 222 to 244 N (50 to 55 lbs) on dry clay or sandy loam and 289 to 311 N (65 to 70 lbs) on concrete.

engine overload and resulting in a waste of fuel without supplying the extra pull needed. In this situation, more weight is needed.

You can tell whether or not you have enough rear-wheel weight by checking the amount of slippage while operating under field conditions (Figure 153). The procedures for checking wheel slippage are given under **Checking and Correcting Tire Slippage**.

Procedures are given here for:

a. Adding Weight to Rear Wheels.
b. Adding Weight to Front Wheels.

a. Adding Weight to Rear Wheels

There are three methods of adding weight to rear wheels. They are as follows (Figure 154):

- **Pumping calcium-chloride solution, or water, into the tires.**
- **Attaching cast-iron weights to the rear wheels.**
- **Using both liquid and cast-iron wheel weights.**

Of the three methods, the most common is the use of liquid in the rear tires. **Calcium-chloride solution** is best. It provides protection against freezing, and the solution weighs up to 30 percent more than water. It does not damage the tube or tire. Its use is approved by the tire manufacturers.

The **liquid should be filled to the level of the upper rim**—about 75 percent liquid filled. Filling to this level is common practice because there is still enough air in the tire to absorb shock and avoid tire bruises. Tests show that tires with this much liquid lose only about 7.5 percent of their resistance to bruising as compared to air-filled tires. Weight is increased in each tire from around 91 kg (200 lbs) with the average size small tire to as much as 250 to 272 kg (550 to 600 lbs) for large tires.

A tire **completely filled with liquid**—100 percent liquid filled—has no air cushion. Since liquid will not compress, tests show that bruise resistance is lowered almost 37 percent.

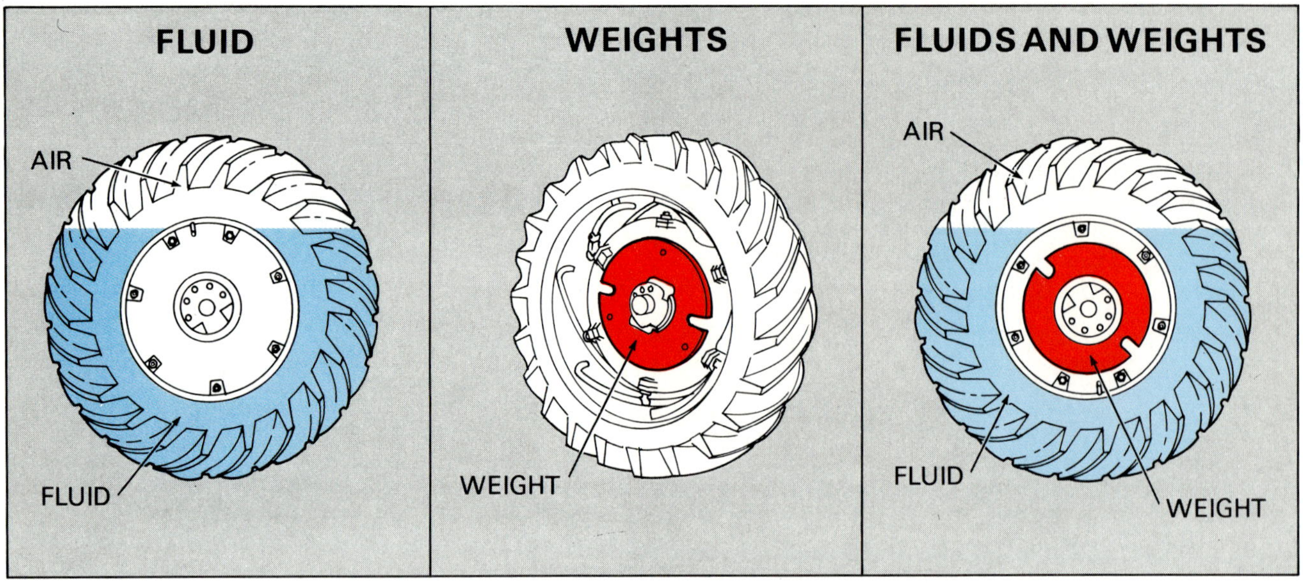

FIGURE 154. Methods of adding weight to rear wheels.

Liquid is in such general use that some tractor manufacturers recommend that the dealer add it to all four tires before delivering it to the purchaser and that it be left in permanently.

This practice is usually best for the owner too, especially if calcium-chloride solution is used. Not many tractor owners are equipped to add solution to their tractor tires. Also, there is a chance of using too much calcium-chloride; under cold conditions the mixture will settle out and not give proper protection against freezing.

For safety and convenience, you should ask your tractor dealer or service station dealer to service your tractor tires with water or a calcium-chloride solution.

Measure liquid height from the lower valve-stem position to the top of the liquid level in the tire. While 3.4 kPa (½ psi) per .305 m (ft) is not the exact figure for either water or calcium-chloride solution, it is close enough to use for either liquid.

Cast-iron wheel weights may be added to each rear wheel and can be used alone or in addition to liquid in the tires. They are usually available in units of 45 to 68 kg (100 to 150 lbs) each. Sometimes they are split to make them easier to mount. As many as two or three weights (or sets of weights) may be added to each wheel (Figure 155).

There is, however, a limit to the amount of weight your tires should carry. Your operator's manual tells you what that limit is for your tractor.

When you are plowing with a moldboard plow and using cast-iron wheel weights, it is good to take one weight from the wheel that runs in the furrow and put it on the wheel that runs on unplowed ground. The reason is that the wheel in the furrow has added weight due to the tilt of the tractor while the wheel on the land is relieved of some weight. By changing the iron weight from the furrow wheel to the land wheel, you equalize traction on the two rear wheels.

FIGURE 155. Two rear wheel weights may be used, or even three, if that much additional weight is needed and if your tires can handle that much additional load.

Manufacturers supply cast-iron weights that fit their brands of tractors. To install, simply bolt the weights into place.

b. Adding Weight to Front Wheels

You can tell whether or not you need more weight on the front wheels by the way they tend to skid sideways while turning with heavy loads. Also, if the front of the tractor tends to lift off the ground when pulling up a slope, you should add front weight.

FIGURE 156. Frame weights added to the front of the tractor.

You can use liquid in the front tires, add cast-iron wheel weights or add frame weights (Figure 156). Or you may use a combination of these.

Weights for front wheels vary from 23 to 30 kg (50 to 65 lbs) each. Frame weights usually run from 45 to 55 kg (100 to 120 lbs) each.

4. ADJUSTING TIRE PRESSURE

You may have to adjust pressure in your tires if you do the following:

- **Add additional weight, such as liquid, mounted equipment, wheel weights or frame weights.**
- **Operate your tractor on paved roads for several hours.**

Even if you check your tires regularly and keep them

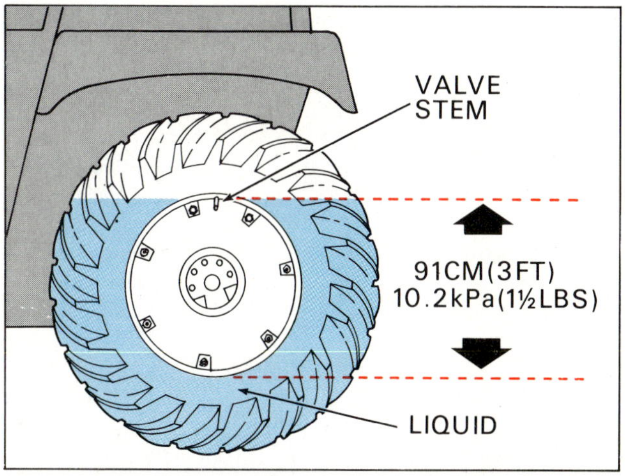

FIGURE 158. Liquid-filled tires can be checked when the valve stem in top position if pressure of liquid is added to air pressure. Add 3.4 kPa (.5 psi) per 30 cm (ft) of liquid height as measured from lower valve-stem position.

properly inflated (Figure 157), it is important to **recheck when weight is added**. Some tractor operators neglect their tires and let the pressure run low much of the time. This habit causes abnormal buckling of the side walls, resulting in cord separation and sidewall cracks.

When checking inflation pressure of liquid-filled tires, turn the wheel so the valve stem is at the top to avoid getting fluid into the tire gage (Figure 158). With the valve in this position, add 3.4 kPa (½ psi) to your reading for each 30 cm (foot) of liquid height in the tire. For example, if the air pressure with the valve at the top is 68 kPa (10 psi) and the liquid level in the tire is 91 cm (3 ft), add 20.6 kPa (1½ psi) of air pressure for liquid pressure to get the actual pressure of 79 kPa (11½ psi).

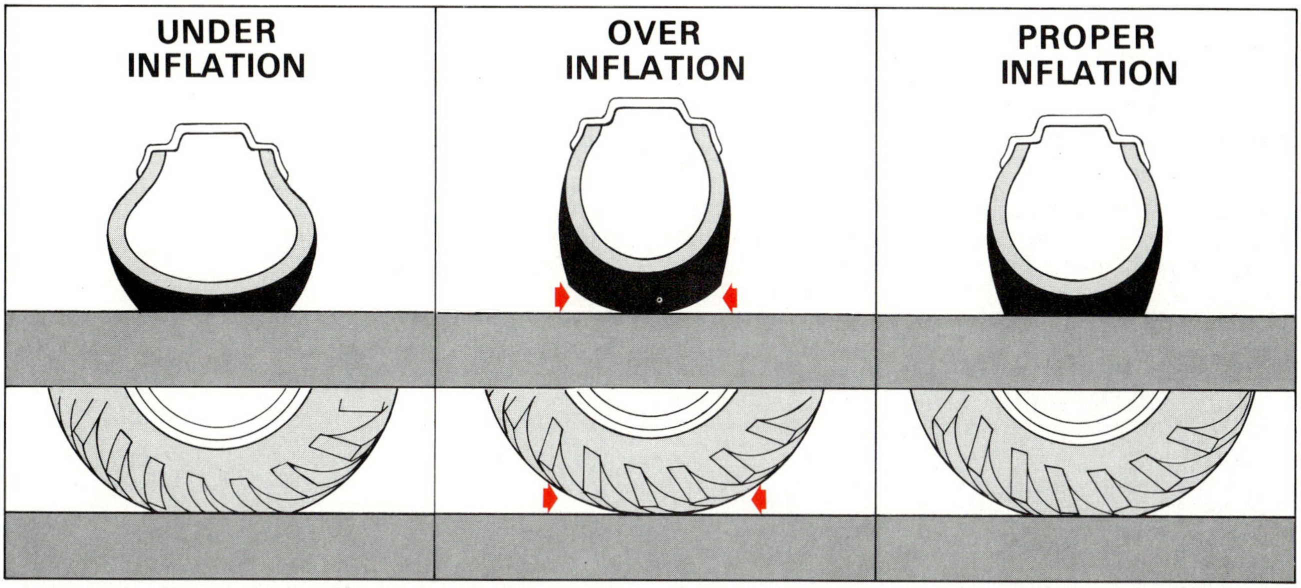

FIGURE 157. Under-inflation causes buckling. It is a common condition with tractor tires. Over-inflation is less common, except when weights or mounted equipment is removed from the tractor without deflating the tires. Proper inflation provides good tire contact with the ground—gives good traction with least wear.

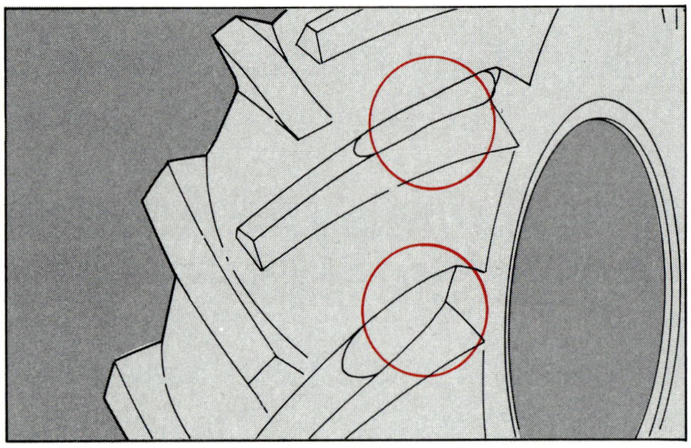

FIGURE 159. Tread-wiping wear due to squeezing and shifting action of tread bars on hard surfaces.

If you plan to operate your tractor on **paved roads** for several hours, it is best to over-inflate the rear tires in order to keep down tread-wiping action caused by the squeezing and shifting of the tread bars as they press and release against hard road surfaces (Figure 159).

If you are **increasing the weight on your tractor tires** or are **planning to operate it on hard surfaces** for several hours, proceed as follows:

1. *Remove valve cap and check pressure with gage.*
   Use a low-pressure tractor-tire gage with one-pound markings to get an accurate check. If you are using water or calcium-chloride solution in your tires, use a gage that can be washed.
2. *Add air (or deflate) as needed to secure proper pressure.*
   Use the tire pressures recommended in your operator's manual. If you have no manual, use Table II as a guide for minimum and maximum pressures and to determine when tires are properly inflated.
   If you are using **mounted equipment** that puts most of the weight on one side of the tractor, put the most pressure in the tire(s) on that side.
   If you plan to operate your tractor on **paved road or other hard surface** for several hours, inflate to maximum pressure (Note Table II).
3. *Wash gage with clear water after using it on tires containing calcium-chloride solution.*
   This procedure is important because calcium chloride will cause rapid corrosion if it isn't completely washed out.
4. *Replace valve cap.*
   The valve core is not always leakproof. Caps help prevent air (and liquid) leaks and at the same time prevent dirt and moisture from entering the valve stems and cores.

TABLE II. PRESSURES FOR FARM TRACTOR TIRES

Developed from American Society of Agricultural Engineers' Standard—ASAE S295-1—adopted June 1965 (reconfirmed 1978), based on standards of the Tire and Rim Association, Inc.

AGRICULTURAL TRACTOR STEERING TIRES (MAXIMUM SPEED 20 MPH)

Tire Size	Ply Rating	Tire Loads at Various Inflation Pressures, lb* 20	24	28	32	36	40
3.00-12	2	175	195	210	225	—	—
4.00-12	4	330	365	400	435	465	495
5.00-15	4	530	590	650	705	755	—
5.50-16	4, 6	655	725	795	860(4)	920	980
6.00-14	4, 6	680	760	830	900(4)	970	1025
6.00-16	4, 6	750	835	915	990(4)	1065	1130
6.50-16	4, 6	860	960	1050(4)	1135	1215	1290
7.50-10	6, 10	825	915	1000	1080	1160(6)	1235
7.50-15	6, 8, 10	1045	1160	1265	1370	1470(6)	1560
7.50-16	4, 6	1110	1235(4)	1345	1455	1560(6)	—
7.50-18	4, 6	1205	1340(4)	1465	1585	1700(6)	—
7.50-20	4, 6	1305	1450(4)	1585	1720	1840(6)	—
9.00-10	8	1080	1200	1315	1425	1525	1620
11.00-16	8, 12	—	—	2515	2720	2915(8)	3100

Tire Size	Ply Rating	Tire Loads at Various Inflation Pressures, lb* 44	48	52	56	60	64
3.00-12	2	—	—	—	—	—	—
4.00-12	4	520	—	—	—	—	—
5.00-15	4	—	—	—	—	—	—
5.50-16	4, 6	1040	1090(6)	—	—	—	—
6.00-14	4, 6	1090	1145(6)	—	—	—	—
6.00-16	4, 6	1200	1260(6)	—	—	—	—
6.50-16	4, 6	1365(6)	—	—	—	—	—
7.50-10	6, 10	1305	1370	1435	1500	1560(10)	—
7.50-15	6, 8, 10	1650	1735	1820(8)	—	—	2055
7.50-16	4, 6	—	—	—	—	—	—
7.50-18	4, 6	—	—	—	—	—	—
7.50-20	4, 6	—	—	—	—	—	—
9.00-10	8	1715	—	—	—	—	—
11.00-16	8, 12	3280(10)	3450	3615(12)	—	—	—

NOTE: Numbers in bold type denote maximum recommended load.
*Based on The Tire and Rim Association, Inc. standards.

AGRICULTURAL TRACTOR DRIVE TIRES (MAXIMUM SPEED 20 MPH)

Tire Size	Ply Rating	Tire Loads at Various Inflation Pressures, lb* 12	14	16	18	20	22	24	26	38
9.3-24	4	965	1055	1140	1220	1300	1370	—	—	—
9.5-24	4	1215	1180	1435	1535	—	—	—	—	—
11.2-24	4, 6	1475	1610	1735(4)	1860	1980	2090	2200(6)	—	—
11.2-28	4, 6	1575	1720	1860(4)	1990	2120	2240	2360(6)	—	—
11.2-36	4, 6	1775	1940	2100(4)	2250	2400	2530	2665(6)	—	—
12.4-16	6, 8	1355	1480	1600	1710	1830	1930(6)	2030	2130	2320(8)
12.4-24	4, 6	1770	1935(4)	2090	2240	2390	2520(6)	—	—	—
12.4-28	4, 6	1890	2070(4)	2230	2385	2550	2685(6)	—	—	—
12.4-36	4, 6	2120	2320(4)	2510	2685	2860	3020(6)	—	—	—
12.4-38	4, 6	2185	2390(4)	2580	2770	2950	3110(6)	—	—	—
13.6-26	4, 6	—	2350(4)	2540	2730	2900(6)	—	—	—	—
13.6-28	4, 6	—	2430(4)	2620	2820	3000(6)	—	—	—	—
13.6-38	6	—	2810	3040	3260	3470	—	—	—	—
14.9-24	6	—	2700	2910	3125	—	—	—	—	—
14.9-26	6, 8	—	2790	3010	3230(6)	3440	3635	3825(8)	—	—
14.9-28	6	—	2870	3100	3320	—	—	—	—	—
16.9-24	6	—	—	3560	—	—	—	—	—	—
16.9-26	6	—	—	3680	—	—	—	—	—	—
16.9-28	6	—	—	3800	—	—	—	—	—	—
16.9-30	6	—	—	3920	—	—	—	—	—	—
16.9-34	6	—	—	4170	—	—	—	—	—	—
18.4-26	6	—	—	4370	—	—	—	—	—	—
18.4-30	6, 8	—	—	4650(6)	4980	5320(8)	—	—	—	—
18.4-34	6, 8	—	—	4940(6)	5300	5650(8)	—	—	—	—
23.1-26	8, 10	—	—	6600(8)	7070	7540(10)	—	—	—	—
24.5-32	10	—	—	8300	8900(10)	—	—	—	—	—
7.50-16	4	820	895	970	1040	1110	1170	—	—	—
13.9-36	4, 6	—	2640(4)	2860	3060	3260(6)	—	—	—	—
15.5-38	6	—	3160	3420	3660	—	—	—	—	—

NOTE: Numbers in bold type denote maximum recommended load.
*Based on The Tire and Rim Association, Inc. standards.

IV. Driving Tractors without Equipment

When driving a tractor regularly, you develop habits. They may be good ones or bad ones. Whether good or bad, they are the ones you will use in any emergency when you haven't time to think. Consequently, it is important that you develop the right habits when you start to learn tractor driving.

From your study of this section, you will be able to start and stop the tractor movement and drive the tractor safely.

This discussion covers basic procedures for operating a tractor without connected equipment. (Operating a tractor with equipment is discussed later.) The information is included under the following headings:

A. Providing for Safety.
B. Starting and Stopping Tractor Movement.
C. Maneuvering the Tractor.

A. Providing for Safety

Safety has been emphasized throughout this program. Now that you are preparing to start driving the tractor, there are additional safety factors to be considered.

Tractor and machinery-related accidents account for almost half of all farm and ranch accidents. Tractors are the primary source. Upon completion of this section you will be able to provide the necessary safety precautions.

The following is a list of "do's and "don'ts" to observe while driving the tractor:

1. *Be sure you are in good physical condition.* If you are sick or exhausted, you should not attempt to operate your tractor. Your chances of error are greater (Figure 160).
2. *Study your operator's manual (Figure 161).* See what it says about driving the tractor.

FIGURE 160. A serious accident may be caused by driver fatigue.

FIGURE 161. Know what your operator's manual says about driving your tractor safely and efficiently.

FIGURE 162. Learn the basic hand signals and teach them to your fellow-workers.

3. *Learn the basic hand signals (Figure 162).*
   If you have someone in the area who may be directing you, be sure both of you understand and use the hand signals.
4. *Check to make sure guards and shields are in place (Figure 163).*
5. *Fasten your seat belt if your tractor has a ROPS (Figure 164).*
   Most tractor accidents are a result of overturning.
6. *Check to make sure the area is clear of obstacles, people and animals (Figure 165).*

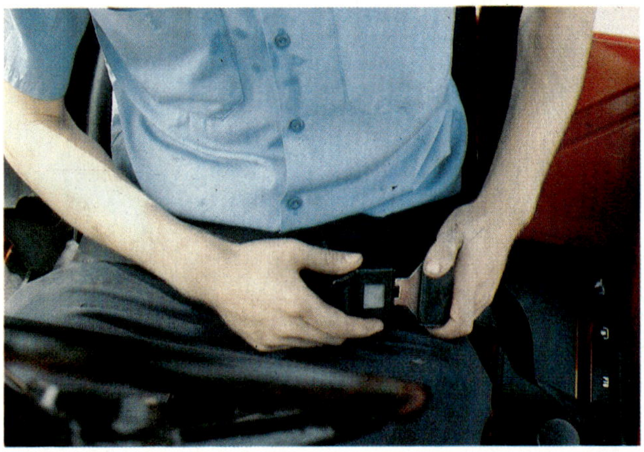

FIGURE 164. Fasten your seat belt if you have a ROPS.

FIGURE 163. Guards and shields must be in place.

FIGURE 165. Make sure area is clear before moving the tractor.

FIGURE 166. No riders.

7. *Do not allow riders on tractor or on trailing equipment (Figure 166).*
8. *Never operate tractor from the ground (Figure 167).*
9. *Do not allow anyone to stand between front and rear wheels on 4-wheel drive tractors (Figure 168).*
   There is a danger of crushing them when the wheels are turned.
10. *Watch out for pinching, cutting and crushing hazards (Figure 169).*
11. *Observe traffic safety rules when driving on roads (Figure 170).*

FIGURE 167. Never operate tractor from any place except the operator's seat.

FIGURE 168. Standing between wheels of 4-wheel drive tractors is dangerous.

FIGURE 169. Beware of pinching, cutting and crushing hazards.

FIGURE 170. Observe traffic rules.

B. Starting and Stopping Tractor Movement

NOTE: Tasks that involve accessories or equipment while the tractor is in motion are not discussed in the following procedures.

Never attempt to start the tractor movement until you are certain you can stop the tractor. Upon completion of this section you will be able to start and stop the tractor properly and safely.

Procedures for starting and stopping the tractor movement are given as follows:

1. Starting the Tractor Movement.
2. Stopping the Tractor Movement.

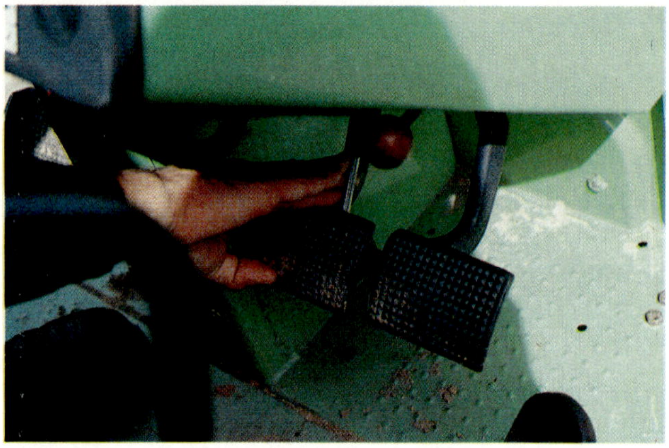

FIGURE 172. On tractors, each rear wheel has a separate brake pedal, but the pedals can be locked together for normal brake operation.

1. STARTING THE TRACTOR MOVEMENT

1. *Adjust the engine speed (Figure 171).*
   On some tractors, the speed is reduced to low idle.

   On others, the speed is advanced slightly.

   Check your operator's manual.
2. *Lock the individual brake pedals together (Figure 172).*
   Tractors usually have a separate brake pedal for each rear wheel. On some, there is one on each side of the tractor. On most tractors, the brake pedals are placed side-by-side. These have a provision for locking the two together. When locked, they operate as a single pedal to brake both wheels at the same time.

   For most *field operations*, the brakes are operated separately for maneuverability.
3. *Disengage the parking brake latch (Figure 173).*
   Hold the tractor with the foot brake if necessary.
4. *Advance throttle slightly.*

FIGURE 173. Before starting the tractor movement, disengage the parking brake latch.

5. *Disengage the clutch and depress the foot brakes (Figure 174).*

FIGURE 171. Adjust engine speed.

FIGURE 174. Clutch in disengaged position.

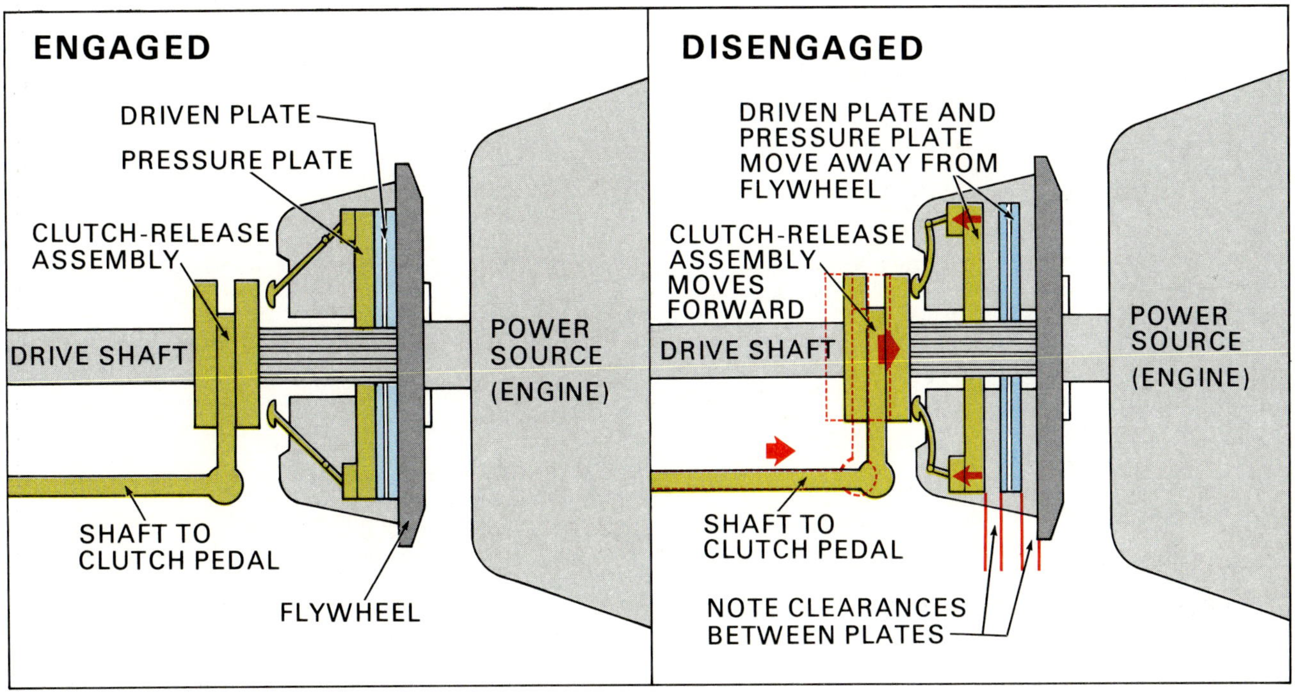

FIGURE 175. Clutch operating principle.

If the engine is running, disengage the clutch by pressing the pedal down with your foot if the tractor is equipped with a manual transmission.

Disengaging the clutch cuts off power to the transmission shaft and allows the gears to stop rotating (Figure 175). The gears must be stopped on most manual shift transmissions if you are to shift without grinding or clashing them.

6. *Select high or low speed range if equipped (Figure 176).*
   Use high speed for light loads if terrain permits. Use low speed for heavy loads and more power. This speed range is not available on all tractors.

If the tractor you are operating shows the gear selection in regular 1, 2, 3 sequence (Figure 177), you have a power-assist shift transmission on your tractor. The inching pedal is equipped with an **inching pedal** located in the same position as the clutch pedal. It is not used for shifting gears but for creeping speeds when attaching implements, slow starts and emergency stops.

It is not necessary to stop tractor movement before you shift into a faster or a slower gear. However, if you change from forward to reverse or reverse to forward motion, stop the tractor before changing gears to avoid gear damage.

FIGURE 176. Some tractors are equipped with a range lever to provide faster or slower speed selection with your regular gear shift.

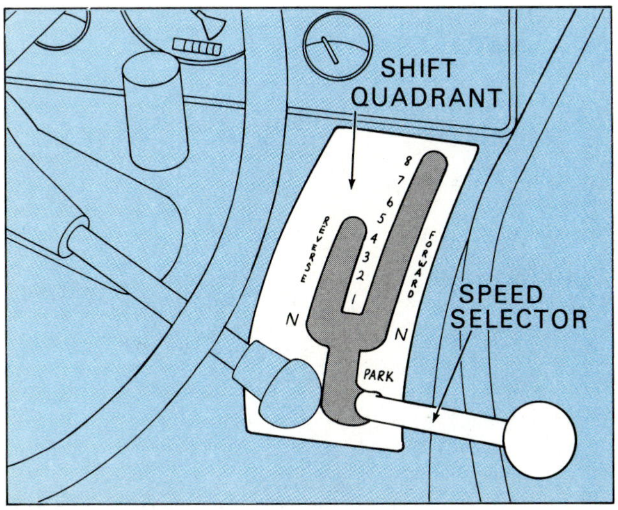

FIGURE 177. Power-assist shift transmissions show the speed selections in regular sequence.

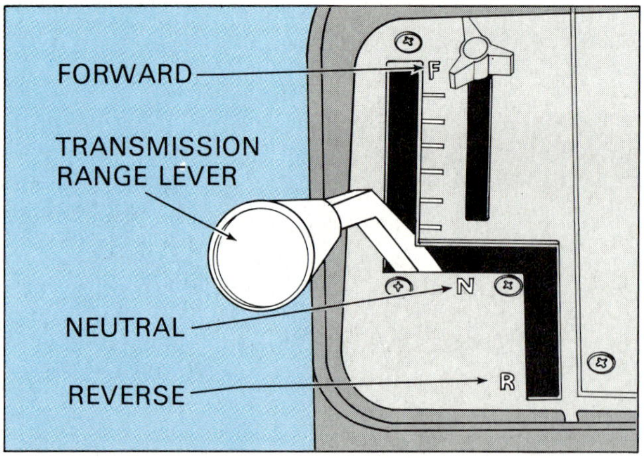

FIGURE 178. Hydrostatic transmissions have no "steps" in speed range selections.

The same is true with a hydrostatic transmission except there are no steps in speed selections (Figure 178).

7. *Move the gear shift lever to the desired gear selection (Figure 179).*
   Select one of the low speeds until you become familiar with the tractor and controls.
   Use low gear selection (first or second) for heavy work and higher gear selection for lighter work.

   NOTE! Follow the manufacturer's recommendations regarding shifting. On most tractors you always start out in the gear in which you intend to drive or operate. If you wish to change gears, you must bring the tractor to a full stop. On other tractors, you may shift while moving under load.

   Small tractors often have 3 speeds forward, but most medium-size tractors have at least 4 to 6 forward speeds. Larger tractors may have as many as 6 to 12 forward speeds.

   Tractors equipped with a wide selection of gear-shift positions usually provide a second lever called a "range lever."

   With the range lever in "low" position, the lowest to highest speed in the 4 gear-shift positions might range from about 2.4 to 12 km/hr (1.5 to 7.5 mph). With the range lever in "high" position, the same gear-shift positions might provide a range of speeds from 4 to 20 km/hr (2.5 to 12.5 mph).

   Always depress the clutch pedal when shifting into or out of gear to prevent possible damage to the transmission.

   If you have difficulty getting into any gear, "jog" the clutch pedal a little to get the gears into a position where they will mesh. Never force the lever. If the gears are not meshed, forcing them will damage the gear parts.

8. *Release the foot brakes and engage the clutch slowly (Figure 180).*
   If you are starting in **high gear**, slow the engine down a little before letting up on the clutch pedal, then increase the engine speed after the clutch takes hold.

   It is essential to smooth starts and your own safety that power be applied gradually to the drive wheels. Do not "pop" the clutch, which means letting the clutch out too fast.

   Some clutches take hold quickly (grab), so it is important that you practice proper clutch use. This is protection for both you and your tractor. Also, if your tractor wheels are frozen to the ground and you engage the clutch quickly, the front wheels may break loose and rise off the ground. In fact, it is possible to turn the tractor over backwards. If the rear wheels are frozen to the ground, shift into reverse to loosen.

   The same situation exists if you engage the

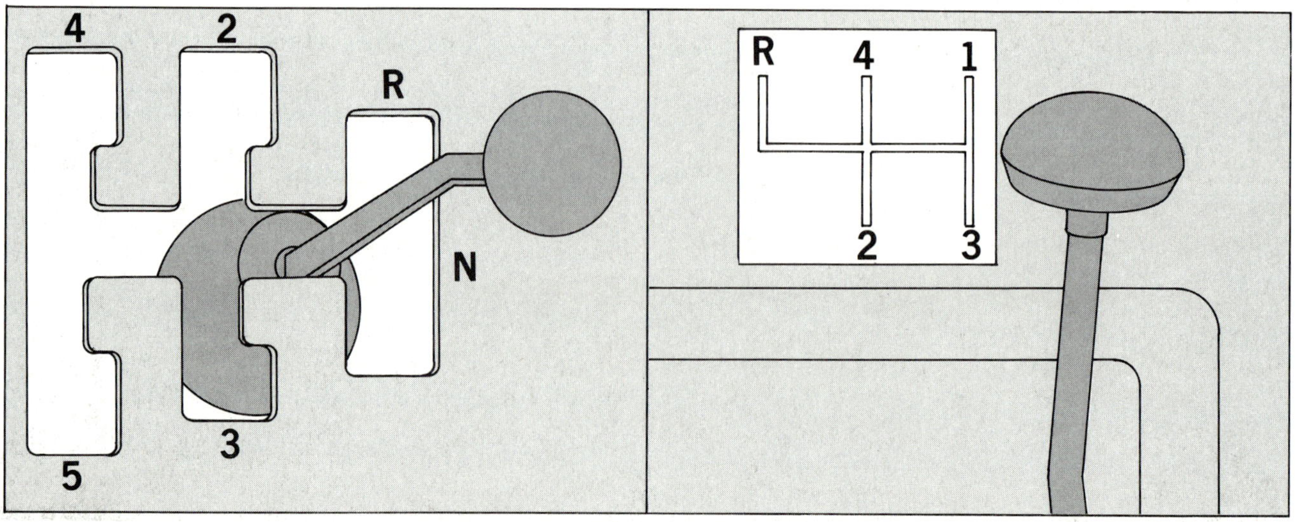

FIGURE 179. Hand-shifted transmissions may have a lever that works within the various slots and notches of a quadrant to provide forward or reverse movement or a lever which is positioned in accordance with a label showing gear positions.

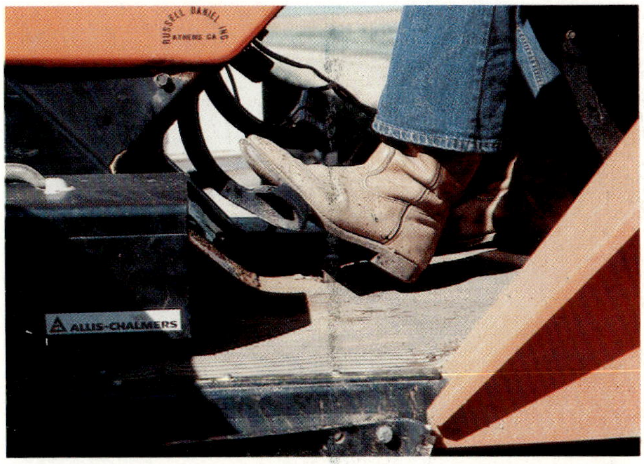

FIGURE 180. Let the clutch out slowly for a smooth start.

clutch quickly with the brakes still locked for parking—the tractor simply rotates around the rear axle.

Manual shift tractors with **planetary-gear transmissions** can be shifted while moving under load. The gears are in mesh constantly. The gear ratios are changed by an internal system of clutches and brakes operated by hydraulic pressure. This arrangement allows you to shift without possible damage to the tractor, but you should do it progressively—one gear at a time.

After the clutch is engaged, completely remove your foot from it. "Clutch-riding," the term used for a person who rests his foot on the clutch during operation, causes the clutch to wear out. Damage shows up in two ways: (1) the clutch facings slip causing fast wear and an early repair job, and (2) the thrust bearing wears rapidly due to oil being wholly or partially scaled off as long as there is some pressure against the clutch pedal.

If your tractor is the type that has an **inching pedal**, don't let your foot rest on it. This habit tends to interrupt power flow and increase wear.

9. *Gradually open throttle to desired speed.*
10. *Shift gears.*

    On tractors with **sliding gear transmissions**, the tractor must be stopped completely before shifting gears.

    On tractors with **power shift and hydrostatic transmissions**, you may shift on-the-go except for changing directions (forward or reverse). To change directions, stop the tractor.

    On tractors with an inching pedal, use it for the slow starts and sudden stops.

2. STOPPING THE TRACTOR MOVEMENT

To stop the tractor movement, proceed as follows:

FIGURE 181. Reduce throttle and apply brakes gradually.

1. *Start preparing to stop far in advance and look for possible hazards.*
2. *Avoid sudden stops whenever possible.*

   Sudden stops could cause the following to happen: the tractor may slide or skid, or the load may jack-knife, which could cause you to lose control and upset the tractor.
3. *Reduce the throttle and apply the brakes gradually (Figure 181).*

   Try to stop on level, solid ground. Leave the tractor in gear and do not disengage the clutch until just before coming to a stop.

   With **hydrostatic transmission**, gradually move the control lever to neutral position. The transmission also acts as a brake. Sudden movements of the control lever will skid the tires or put heavy loads on the drive train.

   With **power-shift transmission**, downshift or depress the inching pedal before applying brakes.
4. *Disengage clutch and move gear-shift lever to neutral or park (Figure 182).*

   Do not put transmission in park until the tractor comes to a complete standstill, as doing so may damage the transmission or power train.
5. *Re-engage clutch.*

FIGURE 182. Disengage clutch and move gear-shift lever to neutral or park.

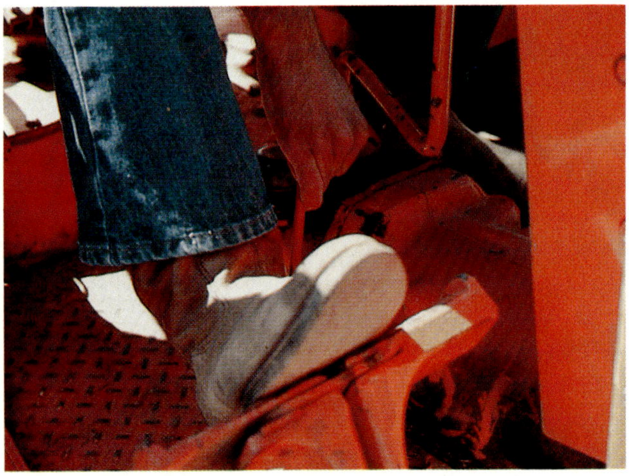

FIGURE 183. Press foot brake down and lock.

FIGURE 184. Be sure to block the wheels if the tractor is on a slope.

6. *Press foot brake down and lock for parking (Figure 183).*
   This procedure is especially important when stopping on a hill or grade. Don't depend on leaving your tractor in gear for holding it on a slope. In the higher gear positions, there is relatively little resistance to movement.
7. *If the engine is to idle, set speed-control lever at fast idle.*
   For carburetor engines, set at 1/4 to 1/3 throttle. For diesels, set at 1/2 throttle.
8. *Allow the engine to idle 2 or 3 minutes.*
   Idling allows the engine to cool.
9. *Stop the engine.*
   If your engine is a **spark-ignition type**, turn off the ignition.
   If your engine is a diesel, close the fuel valve.
10. *Allow enough time for all moving parts to stop before dismounting.*
11. *If you are on a slope, block the wheels of the tractor (Figure 184).*

C. Maneuvering the Tractor

Before starting to handle equipment and loads on a tractor, you should become skilled at driving the tractor without equipment attached.

Operations with which you should be familiar are as follows:

1. Driving on Level Ground.
2. Making Turns.
3. Driving on Streets and Highways.
4. Driving on Rough Ground.
5. Getting Out of a Mud Hole.
6. Preventing Overturns.
7. Operating on Slopes.

1. DRIVING ON LEVEL GROUND

Proceed as follows for

a. Forward Movement.
b. Backward Movement.

a. Forward Movement

Proceed as follows:

1. *Watch the ground ahead of you (Figure 185).*

FIGURE 185. Watch for hidden obstacles ahead of the tractor at all times.

FIGURE 186. Look for people, vehicles or animals.

Look for hidden obstacles that might cause an upset, such as stumps, logs and stones. Look for holes that a wheel might drop into and cause you to lose control. Drive slowly over rough ground and in tall grass and weeds. When mowing, watch for stones or other debris which might be thrown if hit by a blade.

2. *Watch for people, vehicles or animals (Figure 186).*
   Look for any moving vehicles, people or animals that might cross your path and be prepared to stop.
3. *Look ahead for possible obstructions (Figure 187).*
   Look for any obstacles or conditions which might require you to stop or go around. For example, low hanging branches or guy wires from utility poles may require that you go around or reduce speed. Watch for fences, trees, shrubs, flower beds, street curbs or ditches.

FIGURE 187. Watch for obstructions ahead of the tractor.

FIGURE 188. Look behind you and be sure the way is clear when backing.

b. Backward Movement

When backing the tractor, use caution (Figure 188). Some lawn and garden tractors and most industrial tractors have a forward/reverse lever that will reverse the tractor movement with only a very short pause. Be sure to bring the tractor to a full stop and allow yourself time to look where you are going. Follow these steps to back safely:

1. *Stop the tractor.*
2. *Look behind you and be sure the way is clear.*
3. *Depress the clutch and shift to reverse.*
4. *Back at a slow speed.*
5. *Never back any farther than necessary.*
   Turn the tractor around for longer distances except on very steep hills

2. MAKING TURNS

Steering the tractor is about the same as steering an automobile. Many tractors are manually steered, but on some models **power steering** is standard equipment; on others it is optional.

The hydraulic mechanism lessens the amount of pull you exert on the steering wheel to turn the front wheels. With heavy tractors, or when tractors are operating on soft ground or in close quarters, the steering effort is greatly reduced.

Steering is also assisted by the use of the **individual wheel brakes**. For example, if you wish to make a short right turn, by pressing on the right wheel brake the right wheel can be slowed, or stopped, so the front of the tractor turns with little or no forward movement (Figure 189). This is excellent assistance for short

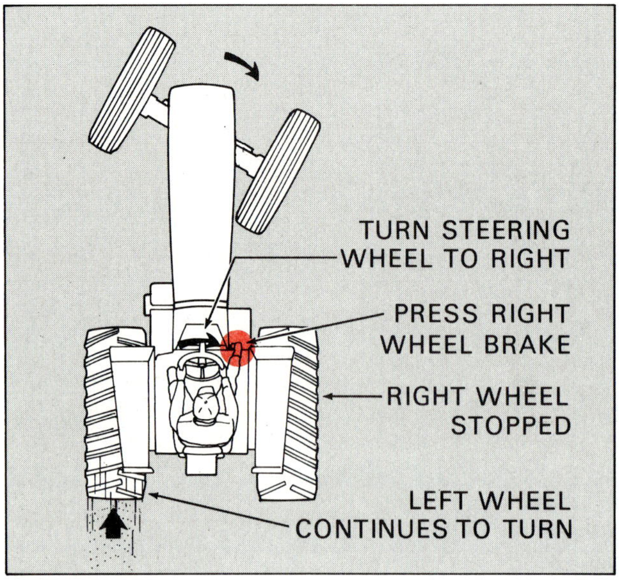

FIGURE 189. Braking one wheel enables you to make a very short turn.

turns at the ends of cultivated rows or when operating a tractor in close quarters such as in a barn or barn lot.

Keep your speed down when making turns. You can upset a tractor at 13 km/hr (8 mph) or less on level ground. Some tractors will travel more than twice that fast. At 26 km/hr (16 mph), your chances of overturning are more than twice as great.

You should have the brakes adjusted so that the same amount of pedal travel will give equal braking on both wheels; then interlock them when traveling at higher speeds. If the brakes are not properly adjusted and locked together and you **press on one wheel brake to make a quick turn, you will probably upset the tractor**. The reason is that when you brake one rear wheel, the other speeds up. If you completely brake one pedal, the other turns twice as fast, greatly increasing the chance of your overturning the tractor.

Next, suppose the inside rear wheel strikes an object on a fast turn (Figure 190). Safety demonstrations conducted by Iowa State University found that ". . . with rear wheels in the narrow position the tractor overturns while turning a 12-m (40-ft) circle at a speed of about 10 km/hr (6 mph) when the inside rear wheel comes in contact with a 30-cm (12-in) block of wood." These are among the unexpected situations which make high-speed turns dangerous.

There is always a possibility that people, pets, vehicles or trees and shrubs might be in the vicinity of your tractor. Therefore, precautions must be observed as follows:

1. *Reduce travel speed when making a turn.*
2. *Look in all directions (Figure 191).*
3. *Signal, if operating in an area that would require it (Figure 192).*
4. *Make the turn when the way is clear.*

FIGURE 190. Even at relatively slow speeds, running over a hidden object can cause a tractor to tip sideways.

FIGURE 191. Be sure to look in every direction.

FIGURE 192. Use standard signals to indicate turns.

3. DRIVING ON STREETS AND HIGHWAYS

Street or highway travel is hazardous for a tractor. Even though you have the right to use the highway with these vehicles for short distances, you stand a good chance of being involved in an accident when you do. For your own safety and the safety of other highway users, stay off the main traveled portion of the highway whenever possible. When you must use the highway, be sure your vehicle complies with all safety requirements and that you observe all traffic laws.

Federal law requires that any vehicle which ordinarily travels under 40 km/hr (25 mph) must display a Slow Moving Vehicle (SMV) emblem while on a public highway. The emblem should be displayed .6 to 1.8 m (2 to 6 ft) above the ground and the point should be up.

In addition to the Slow Moving Vehicle emblem, a red or amber flashing light is required by a number of states (Figure 193). Some states require only that the light be used when visibility is poor. Check the requirements for your state.

FIGURE 193. A flashing amber light will help prevent accidents.

In hilly areas, mount a red or orange flag on a mast higher than the tractor to increase its visibility (Figure 194).

To drive a tractor on streets or highways, proceed as follows:

1. *Take care when entering the highway (Figure 195).*
   Be sure there are no vehicles coming for a long distance.

   Suppose a vehicle is traveling 87 km/hr (55 mph) and you pull out on the highway when the vehicle is 100 m (330 ft) away. The driver of the vehicle has less than 4 seconds to bring his vehicle under control. It will take the vehicle at least 61 m

FIGURE 194. A red or orange flag on a mast helps increase its visibility.

   (200 ft) to make a full stop. If the driver does not see you for one second or if his reactions are not quick, he will hit you from the rear.

2. *Adjust travel speed.*
   Many items should be considered when deciding your tractor speed. They are the condition of the road, the tractor, the traffic and the weather, as well as the kind of a load being pulled, if any.

   Tractors travel too slowly to be safe in automobile traffic. However, tractors can travel too fast for the safety of the tractor and its operator.

   As speed doubles, your chances for upset are four times as great. Use good judgment and adjust travel speed to conditions. Do not drive half on the road and half on the shoulder. Drive completely on the shoulder whenever possible (Figure 196).

FIGURE 195. Do not pull out onto highway until the way is clear.

FIGURE 196. Travel on road shoulder whenever possible.

FIGURE 198. Pull off to the side of the road to allow traffic to pass.

3. *Signal all turns and stops.*
   Signal all turns while on the highway (Figure 197). Be sure your signal is visible.

   Never attempt to make sudden stops while traveling at high rates of speed. Plan ahead and anticipate slowing or stopping, then signal.
4. *Allow traffic to pass if it builds up behind you (Figure 198).*
   Many states require that you pull off to the side of the road and allow cars to pass if you are obstructing traffic. It is the safe thing to do whether required or not.

FIGURE 197. Signal all turns with lighted turn indicators or hand signals.

4. DRIVING ON ROUGH GROUND

The most important safety factor when operating your tractor on rough ground is to reduce speed. You are much more likely to upset the tractor if you hit a stone, hole or stump at 9.6 km/hr (6 mph) than at 4.8 km/hr (3 mph). Be especially alert for hidden objects in tall grass or weeds along ditches and fence rows.

5. GETTING OUT OF A MUD HOLE

If you get stuck in mud using a mounted implement, raise the implement and try to back out in your old tracks.

Trying to drive forward is dangerous. If the mud is deep enough, it will keep the wheels from turning. The tractor will rotate up and backwards around the axle in ½ to 1⅓ seconds (Figure 199). There is not time for the operator to jump clear.

Try digging out in front of the rear wheels, shift into first, use moderate engine speed and engage the clutch slowly.

It will probably help to dig out in front of the front tires, too, unless they are already on solid ground.

FIGURE 199. A tractor can turn over backwards in ½ to 1⅓ seconds.

CAUTION! The practice of chaining a log or plank to the rear wheels is **not recommended.** There is too much danger of tilting the tractor or something else going wrong. Get help, if possible. If another tractor is used, hitch to the drawbar of the towing tractor to prevent overturning.

6. PREVENTING OVERTURNS

With tractors of over 15 kW (20 hp) (since 1974) having ROPS and seat belts, you have some protection from overturns. This is no reason for carelessness, however.

1. *Fasten your seat belt if your tractor is equipped with ROPS.*
2. *Stay clear of ditches, embankments and ponds (Figure 201).*
   A tractor being driven out of a ditch may tip backward when power is applied (Figure 202). Sideways upsets happen easily in ditches. A roll bar will stop a tractor from completing its backward overturn (Figure 203). Seat belts will keep the operator in the safe zone.

FIGURE 201. Do not drive too close to an embankment.

FIGURE 202. The tractor may tip backward when being driven out of a ditch.

NOTE! Cross ditches where banks have a gradual slope—or don't cross them!

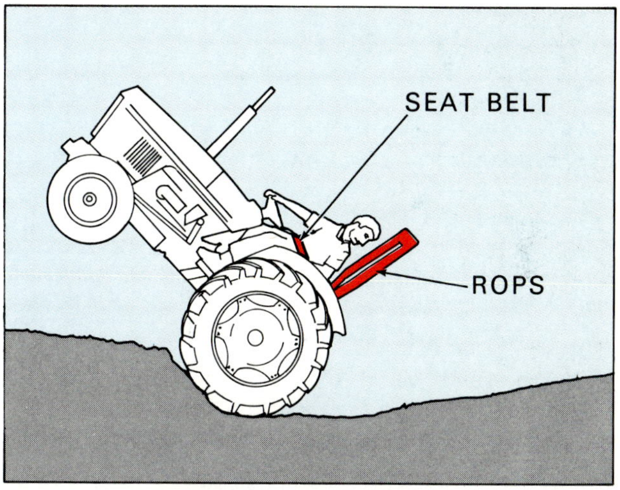

FIGURE 203. Roll bars and seat belts in use prevent many overturns from becoming fatal accidents.

If your tractor starts to tilt while on a slope, turn the steering down hill (Figure 204). This will usually prevent overturn. The natural tendency is to turn uphill, but this will increase the chances of overturning.

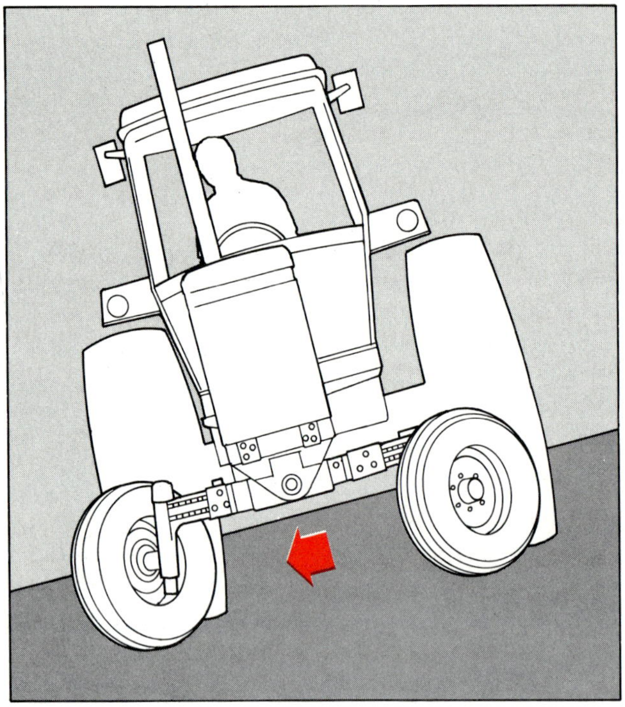

FIGURE 204. Turn the steering down hill if your tractor starts to tilt.

7. OPERATING ON SLOPES

1. *Reduce speed when operating on slopes (Figure 205).*
   A hole, a bump or a quick turn can bring tragedy on a slope. Be extra cautious—if the slope is steep, stay off it, except when operating specially designed tractors with a low center of gravity or a hydraulic levelling mechanism. Use the widest possible wheel spacing on slopes.

FIGURE 205. Stay off steep slopes unless the tractor is designed for them, such as this type.

FIGURE 206. Go up steep hills in reverse.

2. *Do not drive up hills if you can avoid it (Figure 206).*
   Backward upsets can happen when going uphill. If you must climb a steep slope, turn the tractor around and go up in reverse (Figure 206). Again, the exception would be for specially designed tractors. Some models have the center of gravity moved forward and down. Proceed with care even with these tractors on hills.
3. *Leave the tractor in gear when going down hill.*
   This precaution gives you better control of the tractor and the engine acts as a brake.

   Do not go down hill with heavy loads (Figure 207).

FIGURE 207. Do not go down hills with heavy loads.

V. Hitching and Unhitching Equipment

Standards have been established for uniform hitches between tractor makes and models that help to simplify hitching.

There are also many tractors in use that were made before the standards were adopted. Some pieces of equipment can be used on several makes of tractors while others cannot. Check your operator's manual before attempting to attach an implement to a tractor.

From your study of this section, you will be able to **make implement-to-tractor hookups** for **efficient** and **safe operation** of both **implement** and **tractor.**

Hitches are discussed under the following headings:

A. Hitching Equipment.
B. Unhitching Equipment.

A. Hitching Equipment

Hitching equipment is discussed under the following headings:

1. Providing for Safety.
2. Attaching Equipment to the Drawbar.
3. Hitching Rear-Mounted Equipment.
4. Attaching Miscellaneous Equipment.
5. Connecting the Power Take-Off.
6. Connecting the Remote Hydraulic Cylinder.
7. Hitching to Belt-Driven Equipment.

1. PROVIDING FOR SAFETY

Hitching and adjusting implements cause many of the nonfatal accidents. However, they can lay you up during a busy season.

Upon completion of this section you will be able to **provide the necessary safety precautions.**

Accidents are often caused from lifting or being cut, pinched or crushed:

- **Lifting** (Figure 208).
- **Cutting** (Figure 209).
- **Pinching** (Figure 210).
- **Crushing** (Figure 211).

FIGURE 208. When lifting, keep your feet together, bend your knees and do not twist your body.

FIGURE 209. Wear gloves to prevent cutting your hands on sharp objects.

FIGURE 210. Avoid getting your hands pinched.

FIGURE 211. Implements are heavy and can easily crush a person's foot or other parts of the body.

2. ATTACHING EQUIPMENT TO THE DRAWBAR

It is extremely important that any towed equipment be attached to the drawbar (Figure 212). Never attach a rope or chain to the rear axle, seat or to one of the lower links of a three-point hitch. There is grave danger of upsetting your tractor (backwards) or of damaging it mechanically.

FIGURE 212. Use the drawbar when towing equipment.

To attach equipment to the drawbar, proceed as follows:

1. *Check your operator's manual.*
2. *Adjust drawbar height to approximately 38 cm (15 in) clearance between ground level and hitch point on drawbar (Figure 213).*

Three examples of drawbars are shown in Figure 214.

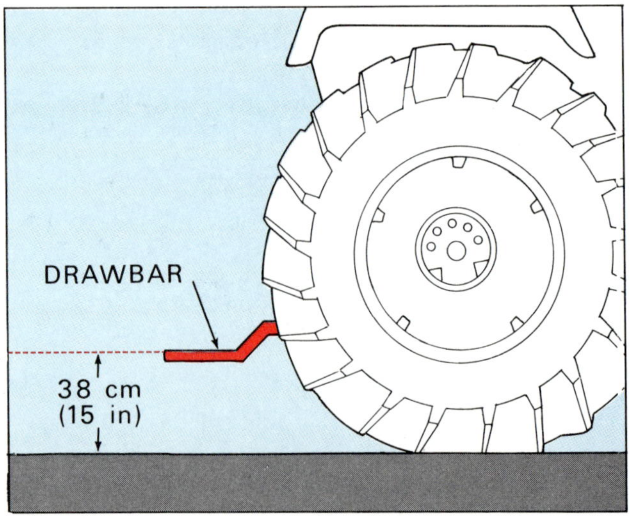

FIGURE 213. A 38-cm (15-inch) drawbar height is commonly used.

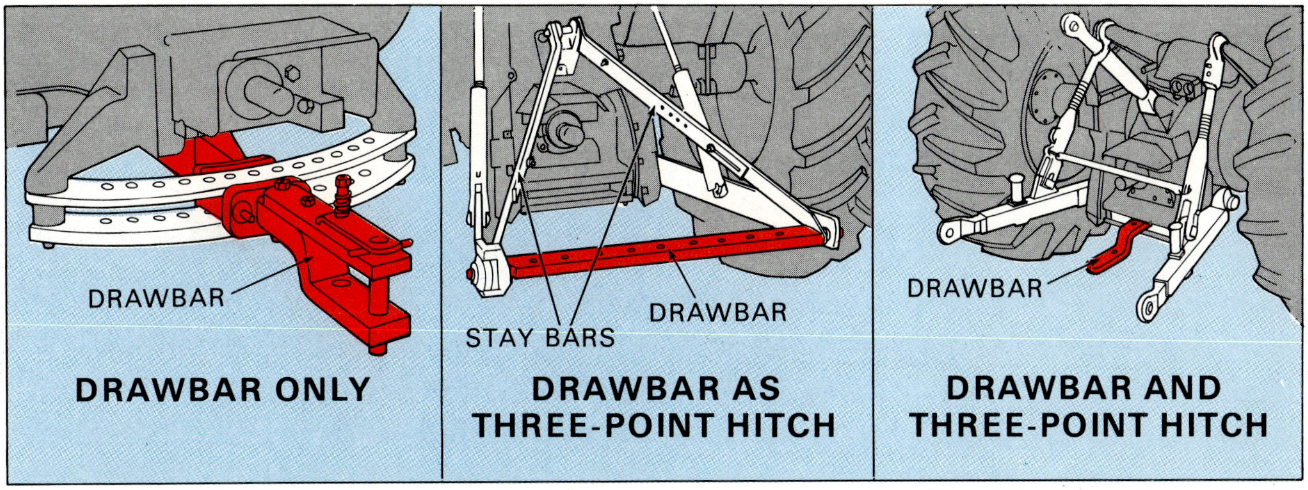

FIGURE 214. A tractor may have a drawbar only, a drawbar mounted as part of a 3-point hitch or separate drawbar and 3-point hitch.

Thirty-eight cm (15 in) is an approximate measurement. If your drawbar is between 32 and 42 cm (13 and 17 in), it is within the standard established by the American Society of Agricultural Engineers and the Society of Automotive Engineers. This not only provides a safe hitching height, but it is also standard height when connecting to a *power take-off shaft*.

Changing height is usually accomplished by (a) turning the drawbar over, (b) adjusting the drawbar bracket, or (c) adjusting the drawbar frame assembly where it bolts to the tractor frame.

If you are using a **drawbar mounted between the lower links of your three-point hitch**, check your operator's manual on how to install **stay bars** or braces to maintain the proper drawbar height. Without them, you might raise the drawbar high enough with the hydraulic control to cause the front wheels to rise off the ground on a pull. Your operator's manual will also instruct you about setting and locking your hydraulic-control lever in a neutral position.

If you are hitching to an implement such as a grain drill or planter and want to keep the seed box level, make the adjustment on the implement hitch if possible.

3. *Check drawbar length setting (Figure 215).* Most tractors provide for a close hitch and an extended hitch. Use the **close hitch** for all jobs except those requiring connection to a power take-off.

   Use the **extended hitch** if you plan to connect to power take-off equipment. There are 2 (ASAE-SAE) standard power take-off speeds—one established in 1923 for 540 revolutions per minute and one established in 1958 for 1,000 revolutions per minute. Most tractor models produced within recent years have provisions for both speeds.

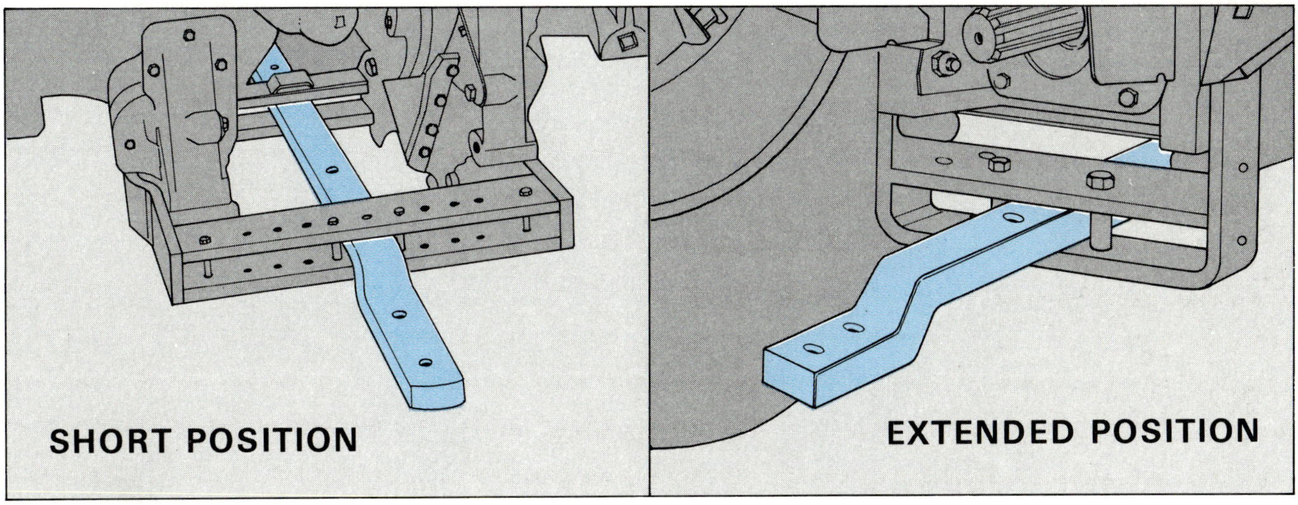

FIGURE 215. Use drawbar in short position for equipment not requiring a PTO connection. Extend drawbar for power take-off hookup.

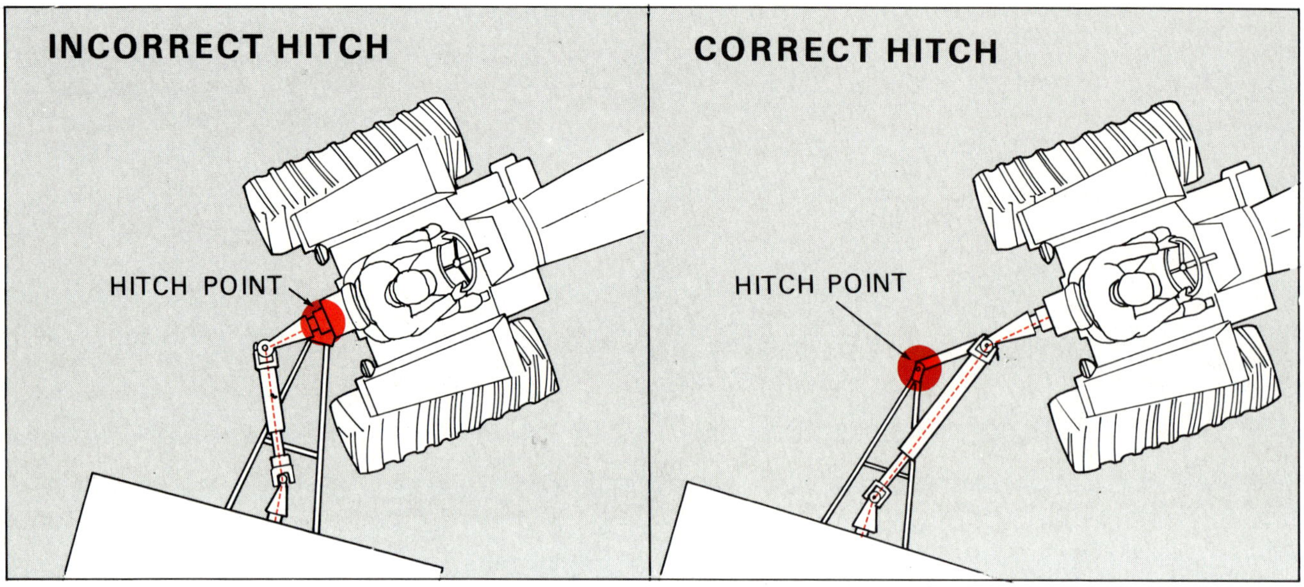

FIGURE 216. With short drawbar hitch, power shaft binds or locks on short turns. With proper drawbar hitch, power shaft can continue to operate on short turns.

For power take-off speed of **540 rpm** (revolutions per minute), the hitch point must be 35 cm (14 in) from the end of the power take-off shaft. For power take-offs operating at **1,000 rpm**, the hitch point must be 40 cm (16 in) from the end of the power take-off shaft.

If your tractor drawbar won't adjust enough to provide a hitch 35 cm (14 in) or 40 cm (16 in) from the end of the power shaft, check with your dealer. He should have a drawbar extension which will provide the proper length. Your drawbar hitch point must be the proper distance from the end of the power take-off shaft. A drawbar connection that is too short causes the first universal joint, back of the power take-off, to lock or bind when you are making a short run (Figure 216). A drawbar of proper length allows the universal joints to turn freely, even on fairly short turns.

4. *Check lateral (sideways) position of drawbar.*

   It is beyond the scope of this discussion to give detailed instructions on how to hitch each item of equipment properly. However, from the standpoint of operating a tractor, these are the drawbar adjustments you can use.

   For **disking, harrowing and pulling most heavy loads**, **let the drawbar swing freely** *(Figure 217)*. It makes the tractor easier to steer, especially where there are short turns to make. Not all tractors have a swinging drawbar. You can tell whether yours is the swinging type or not. Check to see if it can pivot on the opposite end from the hitch; then check the frame section next to the hitch to see if loosening a clamp or removing a bolt or latch will allow it to swing.

   For **power take-off equipment**, center the drawbar hitch point directly under the power take-off shaft and fasten it securely. This is a good way to keep down wear on the universal joints of the power shaft.

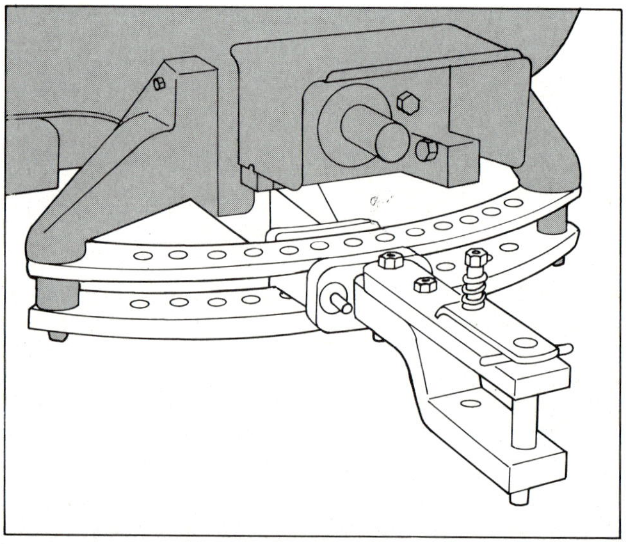

FIGURE 217. Some drawbars are made so that they can swing freely when pulling heavy loads. A swinging drawbar makes a tractor easier to steer.

5. *Back your tractor into position so that hole in drawbar is in line with hole in hitch of implement (Figure 218).*
6. *Set and lock tractor brakes if on sloping ground.*
7. *Attach implement to drawbar (Figure 219).*

   Don't try to position your tractor by operating the transmission controls from the ground. If you do, you run the risk of getting pinned between the tractor and the implement. If you can't insert the hitch pin by moving the implement, get back on

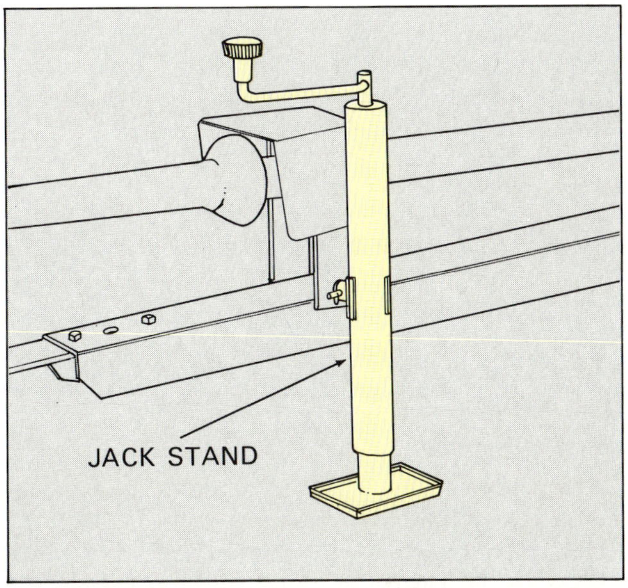

FIGURE 218. Alignment of tractor hitch with implement hitch can often be done from the tractor seat. If not, raise implement hitch to approximate drawbar height with block or with built-on jack or pedestal.

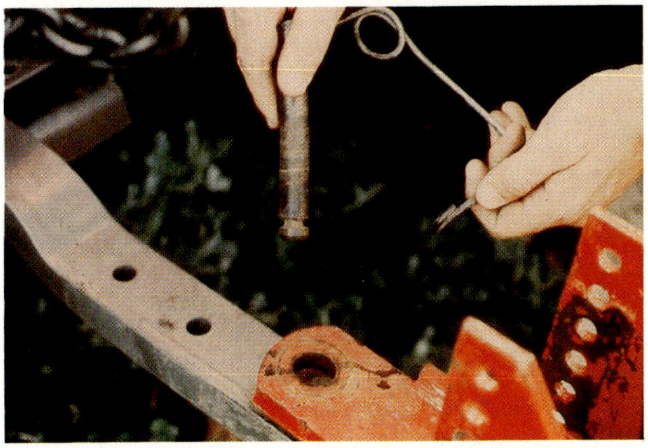

FIGURE 219. Attaching implement to drawbar.

the tractor and realign the drawbar with the holes in the hitch.

The safest way, the one recommended by the National Safety Council, is to lift the implement hitch or drawbar with a steel hook while you are still seated on the tractor (Figure 220). Some tractor operators have equipped the hook with a hitch pin which can also be inserted and latched from the tractor seat.

Be sure you use some means of holding the hitch pin in place so that it won't work out. This precaution is especially important if you are planning to use a power take-off.

8. *Make electrical connections if needed.*
   If you plan to drive on the highway at night, attach the electrical connections for the safety light on your implement.
   Some equipment powered with a separate engine uses an electrical connection from the tractor to provide power for a starter for the implement engine (Figure 221).

FIGURE 220. Aligning the drawbar with the implement hitch.

FIGURE 221. Cable connection between tractor and implement for a safety light or for starting an engine on equipment operated by a separate engine.

3. HITCHING REAR-MOUNTED EQUIPMENT

Equipment that mounts on a tractor has become more popular with tractor owners as improved hitches and lifts have been developed. At first, mounted equipment was lifted and lowered with **hand levers**. They were followed by **tractor-powered gear mechanisms** and **pneumatic-lift mechanisms**. Many tractors still in use are equipped with one of these lifting devices.

Hydraulic pressure for lifting implements was introduced as early as 1933. Since then it has become generally accepted by manufacturers and tractor owners.

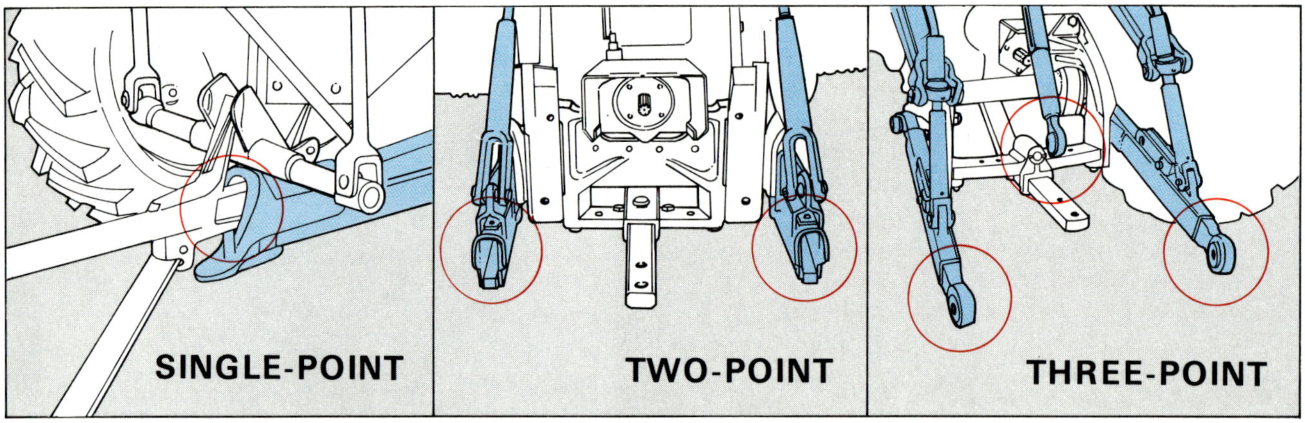

FIGURE 222. Types of integral (rear-mounted, hydraulically controlled) hitches.

At present, all wheel tractors available for farm and industrial use are normally equipped with hydraulic units for controlling equipment. These are called "integral hitches." These hitches are popular because of the ease with which the operator can control equipment.

The various types of integral hitches can be classified into three general groups (Figure 222):

- **Single-point hitch.**
- **Two-point hitch.**
- **Three-point hitch.**

Of the three types, the **three-point hitch** is most commonly used by tractor manufacturers. In fact, it is an exception for a late-model tractor to be sold without having a three-point hitch as either standard or as an option. (Some of the older tractors with other hitch designs can be equipped with adapters that will enable them to use three-point hitch equipment.) The procedures that follow are for the three-point hitch; however, many of the same steps are used in connecting to other hitches.

There are **four categories of implements** designed for the standard three-point hitch. The standard specifications for these categories are as follows:

Category	Maximum Drawbar Power	
	kW	hp
I	15 to 35	20 to 45
II	30 to 75	40 to 100
III	60 to 168	80 to 225
IV	135 to 300	180 to 400

The differences in hitch-point dimensions of the four categories are shown in Figure 223.

Ball Mast Width

Upper Hitch Point Width (Inside)

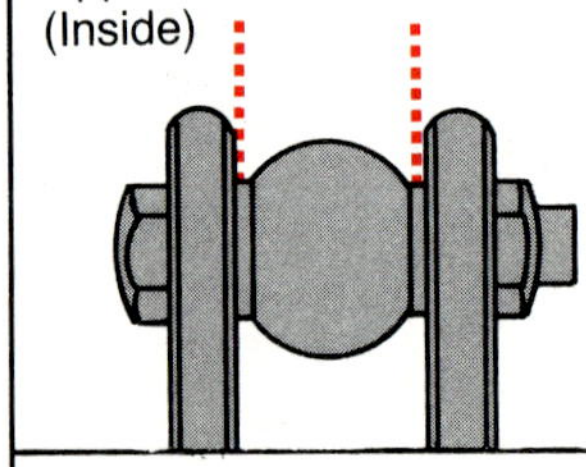

Cat.	cm	in
I	4.4	$1\frac{3}{4}$
II	5.2	$2\frac{1}{16}$
III	5.2	$2\frac{1}{16}$
IV	6.4	$2\frac{1}{2}$

Stud Diameter

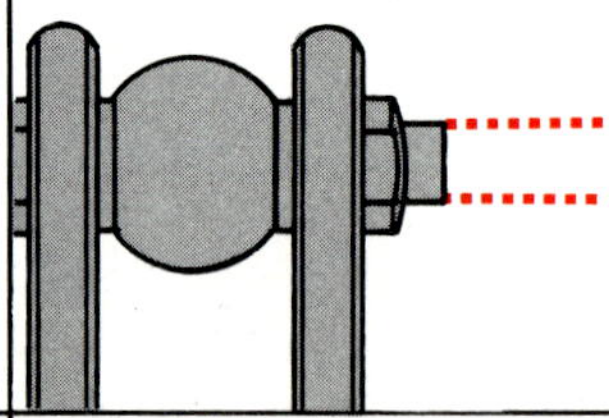

Cat.	cm	in
I	2.2	$\frac{7}{8}$
II	2.8	$1\frac{1}{8}$
III	3.7	$1\frac{1}{2}$
IV	5.1	2

Hitch Pin Diameter

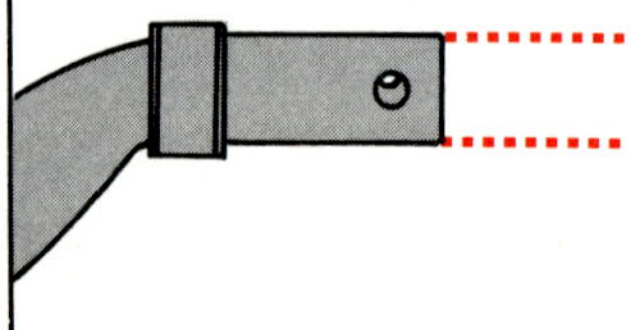

Cat.	cm	in
I	1.9	$\frac{3}{4}$
II	2.5	1
III	3.2	$1\frac{1}{4}$
IV	4.5	$1\frac{3}{4}$

Lower Hitch Spread

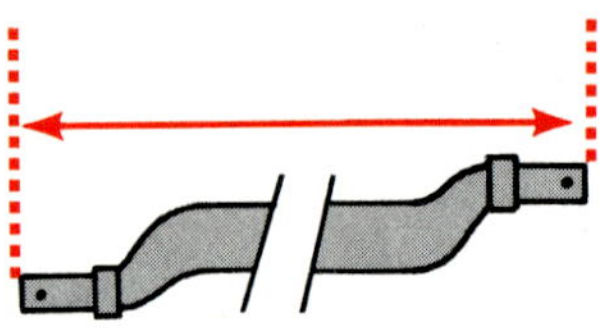

Cat.	cm	in
I	68.3	$26\frac{7}{8}$
II	82.5	$32\frac{1}{2}$
III	96.5	38
IV	116	46

FIGURE 223. Four categories of implement attachments designed for standard three-point hitches. Note differences in measurements.

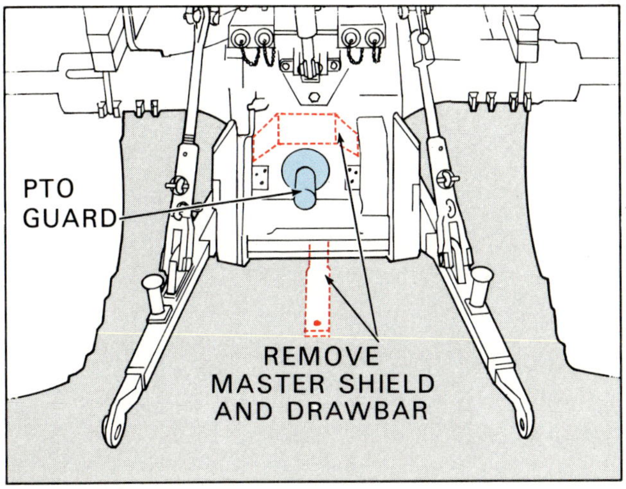

FIGURE 224. With some tractors, the drawbar must be shortened or removed when using the three-point hitch. The master shield over the power take-off shaft may also have to be removed, but be sure shaft is covered by tubular guard, cap or other protection.

To attach an implement to a tractor using a three-point hitch, proceed as follows:

1. *Check drawbar position to make certain it won't interfere with movement of the hitch.*
   Some tractors with 3-point hitches have no separate drawbar. Others, equipped with a drawbar and three-point hitch, must have the drawbar removed for the hitch to work successfully. Still others leave the drawbar mounted, but it must be in the short-connected position (Figure 224) or moved to one side of the drawbar frame.

   With short-coupled implements, it may be necessary to remove the master guard over the power take-off shaft to provide necessary clearance. In that case, **be sure the tubular guard or cap is in position** over the stub power shaft.
2. *Back tractor so lower links are in position for connecting to implement-hitch pins (Figure 225).*

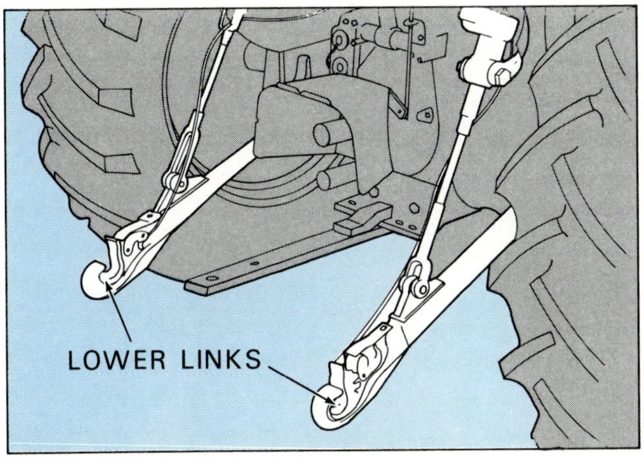

FIGURE 225. Lower links equipped with latches for quick attachment of equipment.

   The term "lower links" has been accepted as standard terminology by the tractor industry and is the term used in this discussion. They do the job of pulling an implement and raising and lowering it. Lower links may also be referred to as "pull links," "draft links" and "draft arms."

   If the lower links are equipped with **ball joints**, back the tractor so that the joints are over the hitch pins on the implement. Since the left link is attached first, its alignment is most important.

   If the lower links are equipped with **latches**, as in Figure 225, align so that latches are under hitch pins of implement.
3. *Raise or lower hitch with hydraulic control to height needed to connect to implement hitch pins.*
   Moving the hydraulic control lever forward or upward (Figure 226) raises the lower links on most tractors and lowers them on others. You can check by moving the lever and watching the movement of the lower links.

   Some implements are connected to a **hitch bar** mounted between the lower links (Figure 227). If you are connecting an implement of this type, back your tractor into the implement hitch. Control the height of the hitch bar so that it will engage the implement hitch.
4. *Place the transmission selection lever in park or neutral.*
   If on sloping ground, set tractor wheel brakes.

   These steps avoid the possibility of the tractor's rolling either forward out of your control or backwards and pinning you between it and the machine.

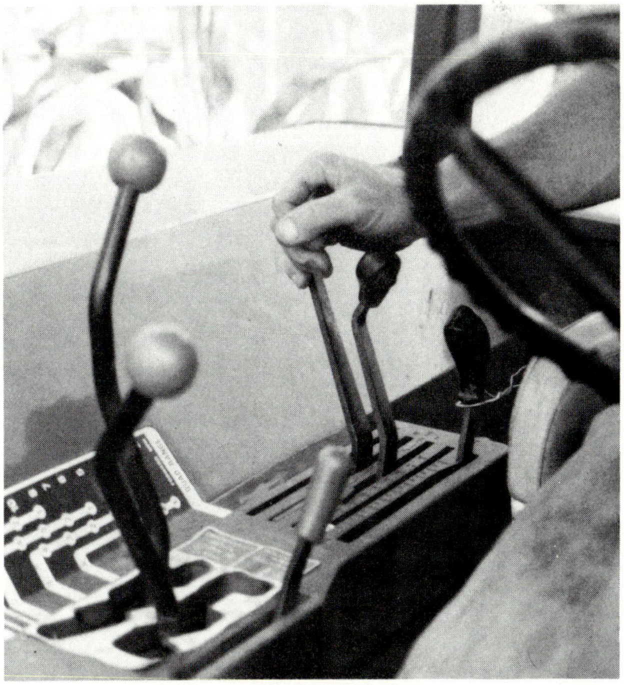

FIGURE 226. Hydraulic-control lever for raising or lowering lower links.

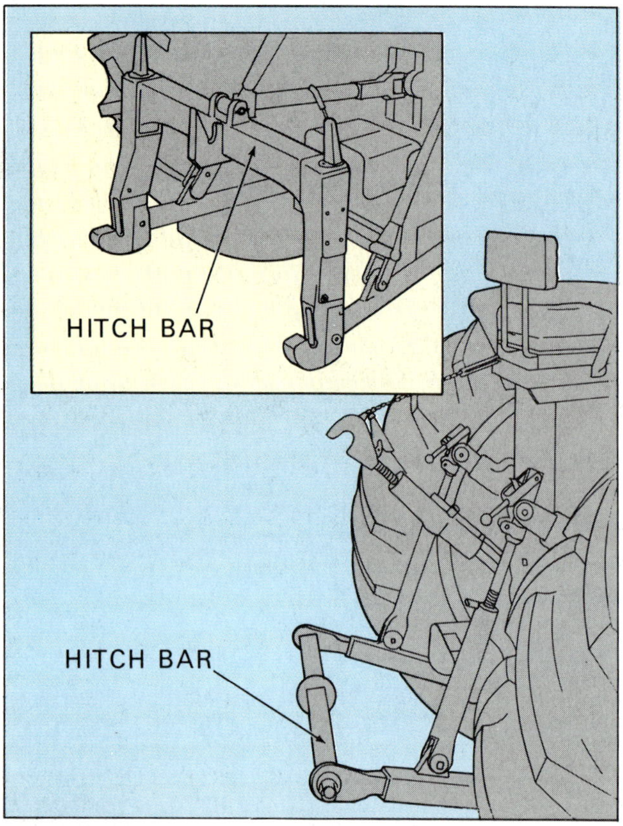

FIGURE 227. Hitch bar used for attaching some implements. (Inset) Quick coupler attachment provided for some tractors.

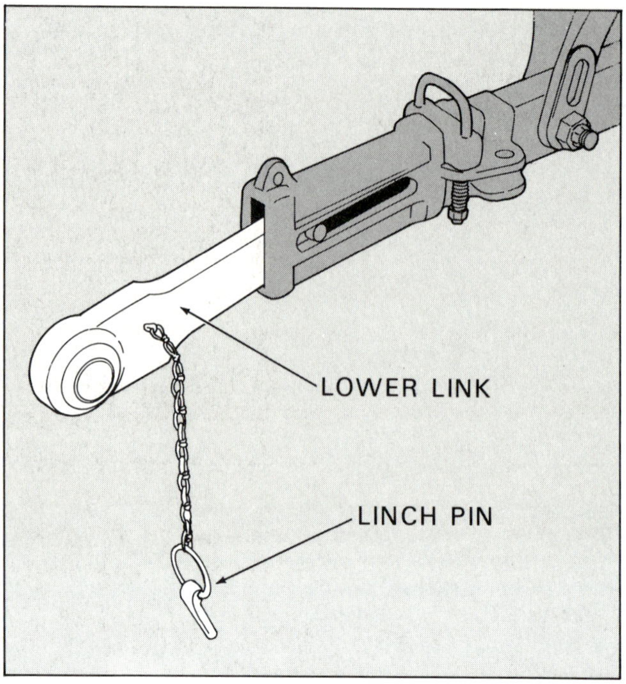

FIGURE 229. Lower link with provision for extending for easy connection to implement.

5. *Attach left lower link to the left hitch pin of implement (Figure 228) and insert linch pin.*

   If lower-link connection is not in line with hitch pin on implement, pull implement ahead, if possible, or rock tractor if it is on level ground.

   If lower-link connection fits loosely on hitch pin, your tractor is probably equipped for connecting the next size category of implement hitches. In that case, use bushings to take up extra space on hitch pins.

FIGURE 228. Attach lower links first.

These are frequently furnished as part of the tractor equipment.

On some tractors, the **lower links can be extended or telescoped** (Figure 229). If your tractor is equipped in this manner, you have several inches of adjustment so it is not necessary to rock the tractor or pull on the implement.

6. *Attach right lower link to implement hitch pin and insert linch pin.*

   Use the same procedures as in step 5 except you can raise or lower the link by turning the crank as shown in Figure 230.

FIGURE 230. Right lower link can be raised or lowered by means of a crank.

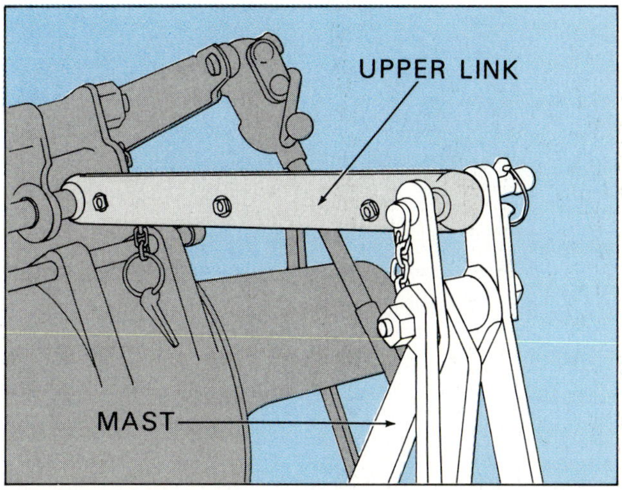

FIGURE 231. Upper link connection to implement mast.

FIGURE 232. Sway locks (stabilizer bars) eliminate swinging action of hitch.

7. *Attach upper link to implement mast.*
   The "upper link" is the third connection between the tractor and implement (Figure 231). This, too, has been accepted as a standard term, but other terms still in use are "turnbuckle link," "top link" and "link-screw assembly."

   It is this adjustment that enables you to tilt the implement forward or backward to level it from front to back.

   On some tractors, the implement connection provided on the upper link is the same as that for the lower links. On others, some type of latch is provided.

   If it is necessary to provide a bushing for the lower link connections, you will also need to change to a larger mast ball or use a bushing in the upper link.
8. *Add sway locks (stablizer bars) if needed.*
   The three-point hitch assembly allows an implement to sway in much the same manner as a swinging drawbar. Action is desirable for such implements as plows, disk harrows and tillers which should be allowed to trail freely. **Sway locks or stabilizer bars** are provided when you wish to stop swinging action for such implements as toolbars, cultivators, bedders and planters (Figure 232).

   Don't confuse sway locks (stabilizer bars) with stay braces or chains. With **sway locks**, you can raise or lower the hitch. Only the swaying action is eliminated. With **stay braces** you can't raise or lower the hitch. It is held rigid.
9. *Raise implement with hydraulic control to transport position to make certain it clears at all points and that the connections are secure.*

4. ATTACHING MISCELLANEOUS EQUIPMENT

In addition to rear-attached and rear-mounted equipment, there are pieces of equipment that are attached to other parts of the tractor.

Front-mounted equipment, such as a cultivator (Figure 233), has not been standardized for all makes of tractors. Most manufacturers have designed and built front-mounted equipment to fit their own tractors of various sizes. Much of this equipment is not interchangeable with different makes.

FIGURE 233. Front-mounted cultivator.

FIGURE 234. Front-end loader.

Center-mounted equipment such as front end loaders (Figure 234) are quite common but like front-mounted equipment are not often interchangeable with most of the different tractor makes.

5. CONNECTING THE POWER TAKE-OFF

Most tractors have a rear power take-off (PTO) shaft (Figure 235). Some have auxiliary PTO shafts that are located toward the front of the tractor (Figure 236).

The **rear PTO shaft** has two speed standards established jointly by the American Society of agricultural Engineers and the Society of Automotive Engineers. They are (1) the original standard of 540 rpm, accepted in 1938 and (2) the second standard of 1,000 rpm accepted in 1958. The 1,000 rpm standard was developed to provide greater PTO efficiency and to better meet the needs of higher speed machines and those with increased power demands.

The standards provide that **shafts shall be splined differently** so that it is impossible to connect a 540-rpm implement to a 1,000-rpm PTO shaft or vice versa.

FIGURE 235. Standard rear PTO shaft.

FIGURE 236. Auxiliary PTO shaft.

Figure 237 shows two stub shafts and the difference in the number of splines on the ends where the implement connects.

A change in PTO speed is necessary if you are preparing to switch to an implement that requires a different speed from that previously used. For example, suppose you have been using a hay baler requiring 540 rpm but want to connect to an irrigation pump that requires 1000 rpm. Then a speed change is necessary.

Two speeds for rear PTO's are provided in three ways:

- **Two permanently installed PTO shafts—one for 540 rpm, the other for 1,000 rpm (Figure 238).**

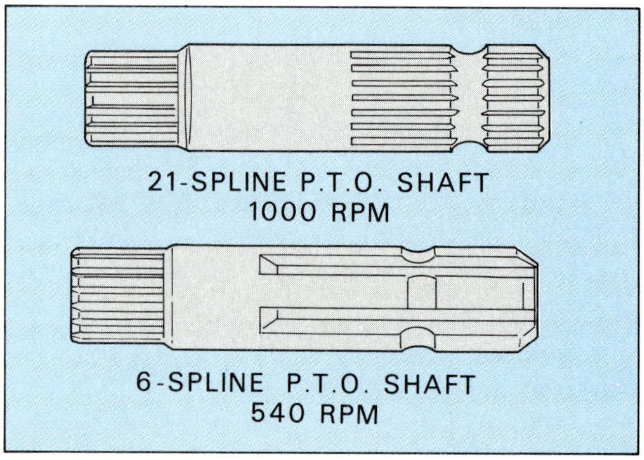

FIGURE 237. PTO shafts for 1,000 rpm have 21 splines on the end where the implement connects. Those for 540 rpm have 6 splines.

FIGURE 238. Two permanently installed PTO shafts—one for each speed.

These are designed so that each is independently driven. Both speeds—540 and 1,000 rpm—are available at all times. Both shafts rotate at the same time. When one shaft is in use, the other must be shielded.

- **Two interchangeable stub shafts (Figure 239).** These are designed so that either shaft can be

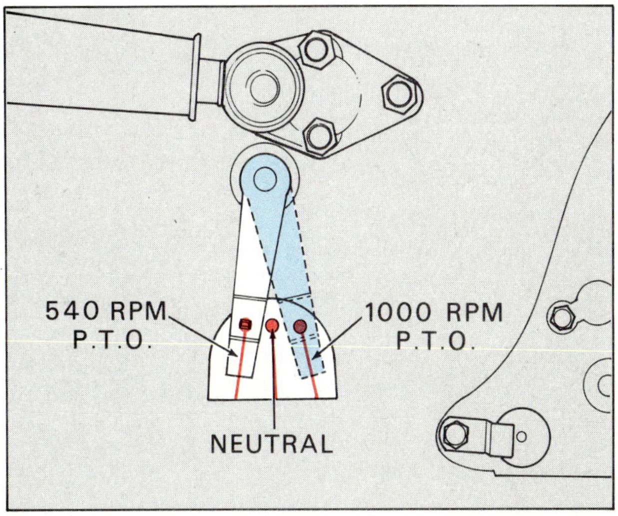

FIGURE 240. Stub shafts can be interchanged on some tractors by using a speed shift lever located on the tractor.

inserted into the PTO outlet. With one type, the speed change is automatic. When the 1,000 rpm stub shaft is inserted, it automatically engages the end of the shaft supplying power directly from the engine. With the 540 rpm shaft engaged, power is diverted from the engine drive shaft through a set of gears to the stub shaft.

Another design of the interchangeable stub-shaft type PTO has a provision for changing speeds by shifting a speed shift lever on the side of the tractor (Figure 240).

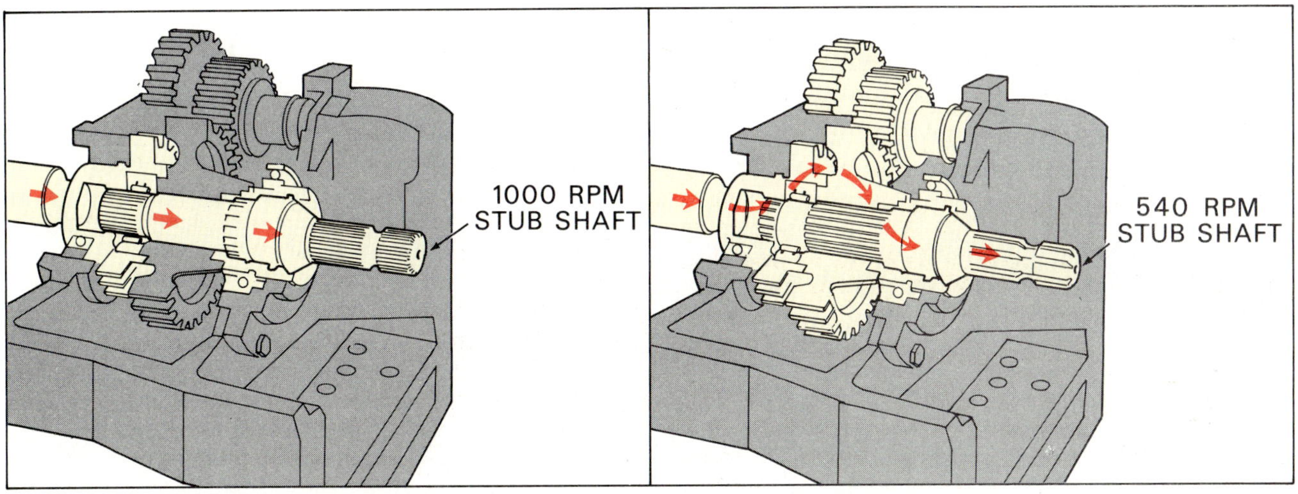

FIGURE 239. Two stub shafts. (Left) stub shaft when installed connects directly to engine shaft to provide 1,000 rpm. (Right) stub shaft engages different set of splines in transmission. Power is automatically diverted through gear train to supply 540 rpm.

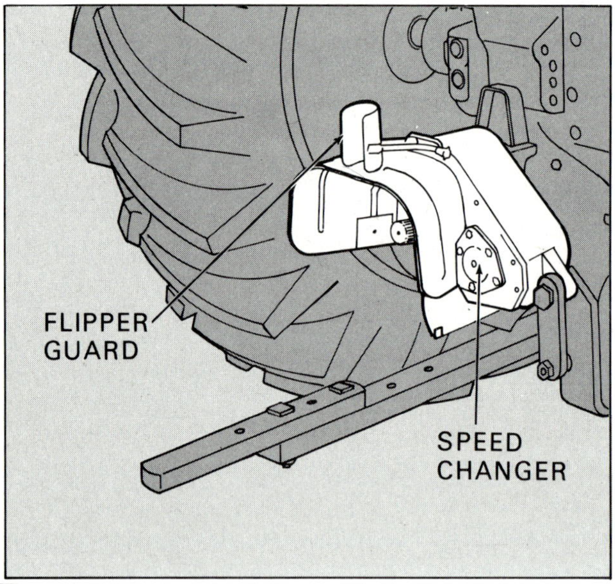

FIGURE 241. Conversion unit to provide 1,000 rpm speed from 540-rpm PTO.

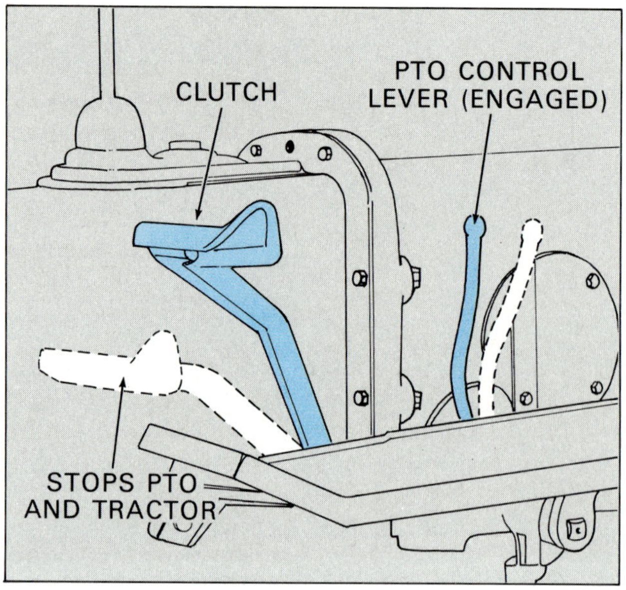

FIGURE 242. Transmission-driven PTO. Clutch engages (or disengages) power shaft and transmission shaft at same time.

- **Conversion units (Figure 241)** used to provide 1,000 rpm from 540-rpm PTO's are available for some tractors. They are not widely used and, over a period of years, will probably be discontinued as newer tractors replace older ones.

Besides the two PTO speeds, there are **three types of PTO drives**. They are as follows:

- Transmission Driven.
- Independent Running.
- Continuous Running.

The **transmission-driven PTO** is the oldest. It has been in use since PTO's were first added to tractors in the 1920's. It is controlled by the main (engine) clutch that controls ground movement of the tractor. With it, the PTO operates whenever you engage the clutch to start tractor movement (Figure 242).

This is a disadvantage when there is a need for a PTO machine to continue running when the clutch is disengaged. This is often the case while combining. The combine may overload, but there is no way to stop forward motion and let it continue to run to clear itself without shifting the tractor transmission to neutral and re-engaging the clutch.

To overcome the disadvantage of the transmission-driven PTO, the **independent PTO** was developed (Figure 243). Each is controlled by means of a separate drive. If you wish, you can continue to operate your machine whether the tractor is in motion or standing still or in the process of being started or stopped.

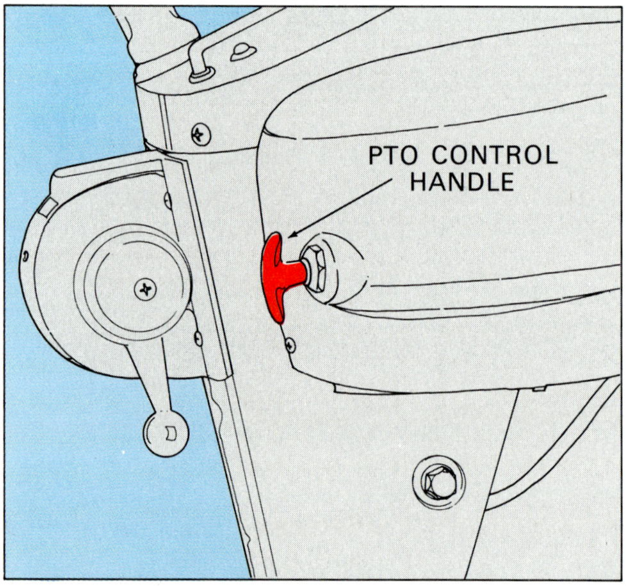

FIGURE 243. Independent power shaft controlled by separate hand control. Power shaft operates independently of ground movement of tractor.

The **continuous running** PTO operates in much the same manner as the independent PTO except that a double transmission clutch is used (Figure 244). The continuous running PTO shaft is controlled by the engine clutch but a second clutch is provided for starting or stopping tractor ground movement.

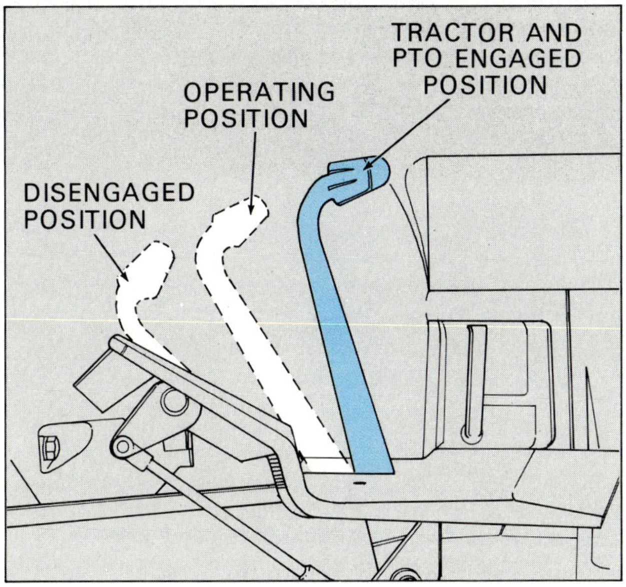

FIGURE 244. The continuous running PTO shaft is controlled by double-acting clutch. With clutch ½ depressed, tractor movement stops but power shaft runs. With clutch completely depressed, both shafts are disconnected. With clutch completely engaged, power is supplied to both shafts.

The discussion that follows is limited to connecting to the standard rear-mounted PTO shaft. However, many of the steps for connecting to the auxiliary PTO shaft are the same. To connect the PTO shaft, proceed as follows:

1. *Set wheel brakes.*
   A rolling tractor can get out of control and easily pin you against the implement that is being attached.
2. *Disengage power to PTO shaft (Figure 245).*
   Disengage with shift lever or clutch or move shift lever to neutral.
3. *Shut off engine.*
   This is good procedure as a precautionary measure and helps form a good habit.

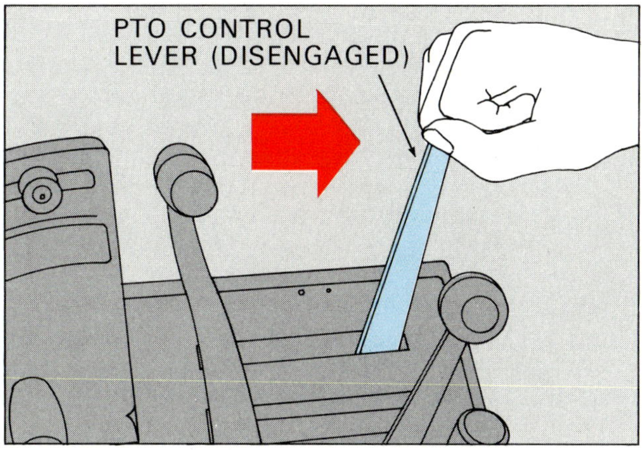

FIGURE 245. PTO control lever disengaged.

FIGURE 246. Removing PTO tube-type guard as installed on some tractors.

4. *Remove cap or tube from end of PTO shaft if installed (Figure 246).*
   Not all tractors are equipped with guards of this type.
5. *Check tractor PTO shaft to determine shaft speed.*
   If the shaft has six splines, it is for 540 rpm. If it has 21 splines, it connects to 1,000-rpm equipment. Both shafts are the same size—3½ cm (1⅜ in) in diameter (Figure 247).
6. *Check splined coupling on equipment to be connected to PTO.*
   If the coupling has the **same number of splines** as are on the PTO shaft, proceed with step 7.
   If the coupling has **more or less splines** than are on the power take-off shaft, proceed as follows:
   (1) *Determine if your tractor has one or two PTO speeds.*
   If you are not certain, check your operator's manual.

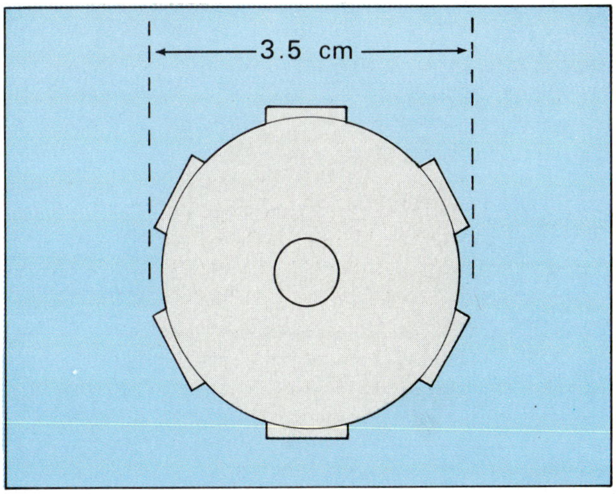

FIGURE 247. PTO shafts are 3½ cm (1⅜ ins) in diameter.

FIGURE 248. Use clean cloth to remove dirt from inside spline of implement drive shaft.

If only one speed is available, it is probably 540 rpm. To connect to the equipment, you would need to add a conversion unit to your tractor. However, this is not a good practice as it would be necessary to run the tractor engine considerably faster or slower to match the speed required by the PTO-driven equipment. In some instances, it may also be necessary to adjust the tractor ground speed. **If two speeds are available**, proceed with (2).

(2) *Exchange stub shafts.*

(3) *If stub shafts are exchanged, make certain shaft speed is correct if changed manually.*

7. *Check inside spline of implement drive shaft for foreign matter or dirt (Figure 248).*
   When last used, the spline may have been dropped on the ground. Dirt, metal particles, etc. stick readily to the inside oily surfaces. If not removed, they cause a grinding and wearing action.
8. *Clean and oil outside spline of tractor PTO shaft (Figure 249).*

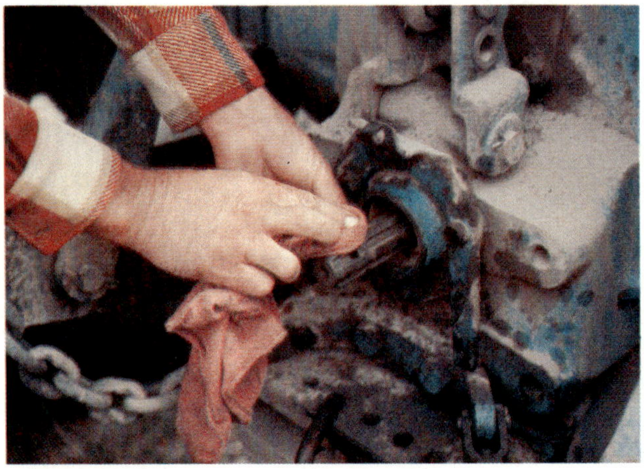

FIGURE 249. Cleaning the PTO shaft.

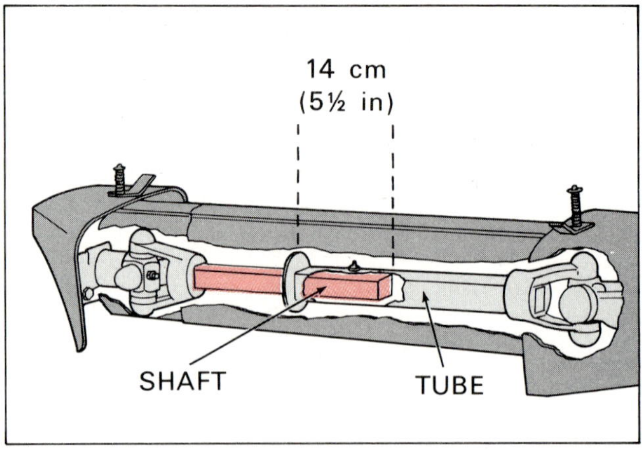

FIGURE 250. Telescoping arrangement on implement drive shaft provides for lengthening and shortening. There should be at least 14 cm (5½ ins) of shaft length extending into tube (with tractor and implement on level ground and both in a straight line) to be certain shafts won't separate when rear wheels of tractor pass over a ridge.

9. *Check telescoping shaft.*
   The implement drive shaft is telescoped to take care of constant lengthening and shortening which occur with each turn or as the equipment passes over unlevel ground. Figure 250 shows the telescoping arrangement between the two universal joints next to the tractor.

   Some of the telescoping shafts are **nonseparable**. This type seldom needs checking. The **separable type** can become so extended that the two sections separate.

   Separation of shafts is dangerous because of the whipping action from that part of the drive shaft still left connected to the PTO. For safety, at least 14 cm (5½ in) of the shaft should be telescoped into the tube after the drive shaft connections are complete. Measurement is determined while both tractor and implement are in a straight line and on level ground.

   If you have a **separable drive shaft**, make certain you have assembled it properly. Most of them are designed with some provision so that you can't make a wrong assembly (Figure 251). If there is no provision of this type, check the positions of the universal-joint yokes at each end of the shaft. Both should be in the same plane. If assembled improperly, the speed of the shaft on the implement end will be irregular, causing severe strain on both the implement and drive shaft.
10. *Slide universal joint into position on PTO shaft and fasten securely (Figure 252).*
   If the universal joint is too large for the tractor PTO shaft, you will need an adapter similar to the one shown in Figure 253. A non-standard size 540 rpm shaft of 2.9 cm (1⅛ in) was used on some of the older-model tractors.

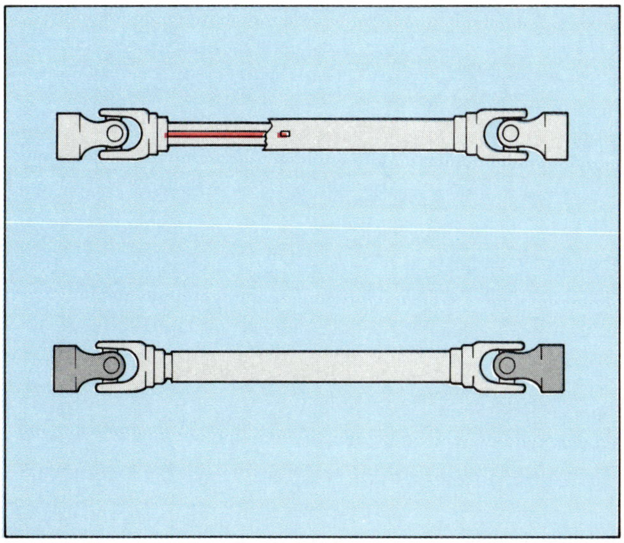

FIGURE 251. Most separable drive shafts are constructed with a bead and groove or other device, to assure correct assembly if they come apart. Shafts without such provision must be assembled so that both end yokes are in the same plane.

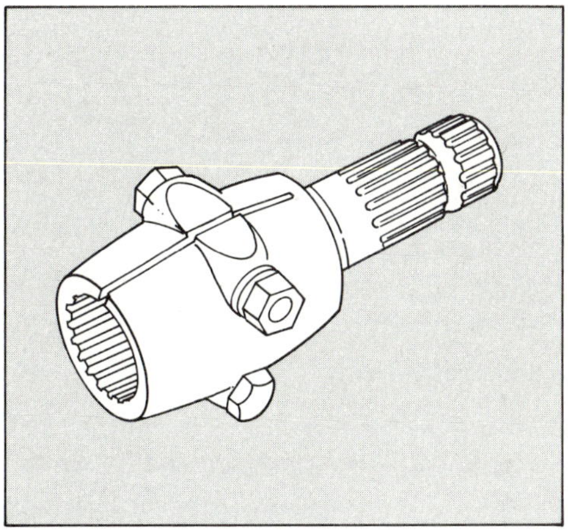

FIGURE 253. PTO adapter.

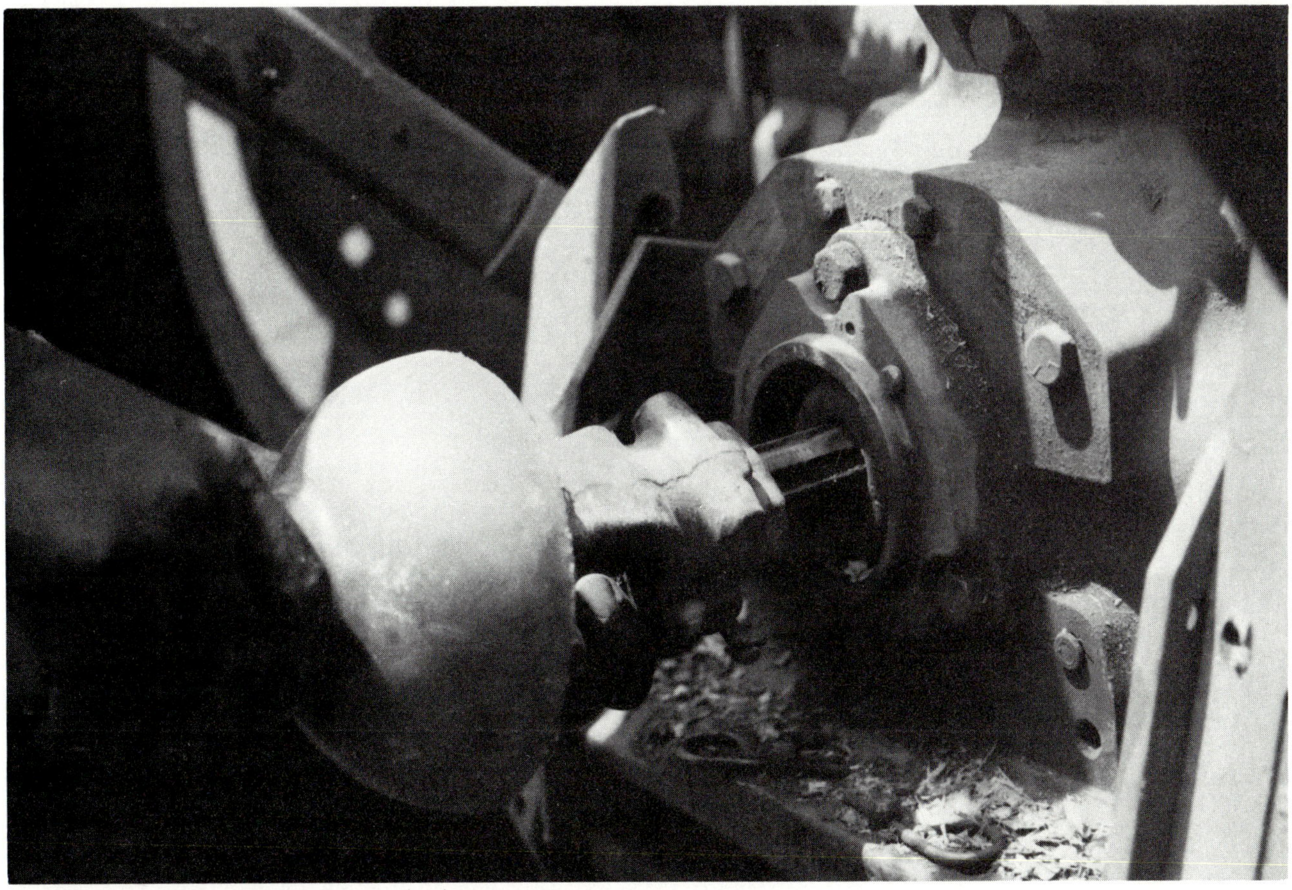

FIGURE 252. Slide universal joint on tractor PTO shaft as far as it will go.

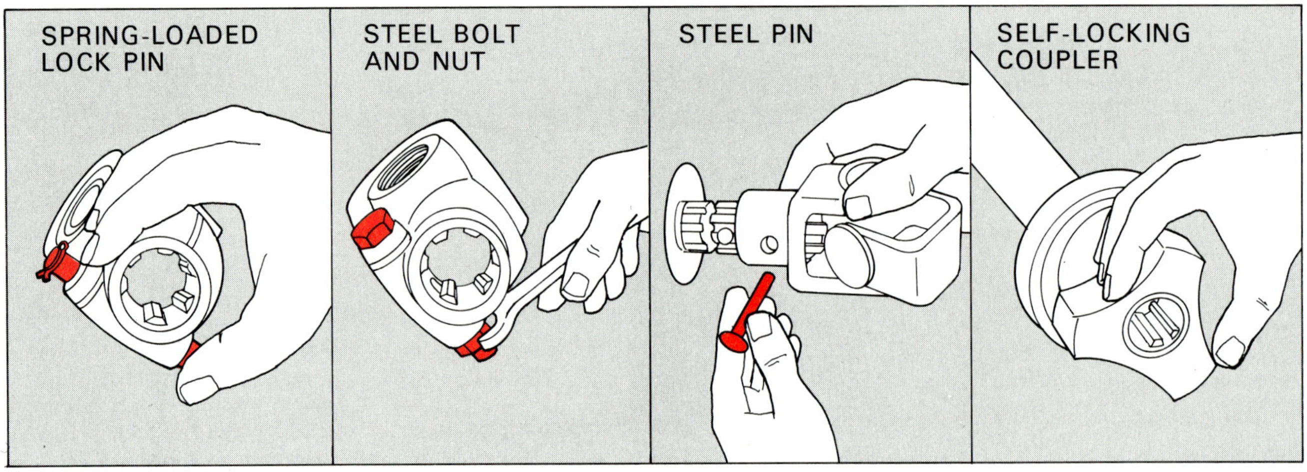

FIGURE 254. Means provided for holding implement power shaft to tractor PTO shaft.

There are 4 means of fastening the universal joint to the tractor PTO shaft (Figure 254). The first two require a grooved shaft which is not on some models of tractors, but all tractor power shafts and universal joints are drilled so that the two units can be pinned together. The fourth type, the self-locking coupler, works much the same as the spring-loaded lock type.

11. *Check connection by pulling firmly on drive shaft assembly.*
12. *Check alignment of PTO shaft assembly to make certain hitch point is centered between universal joints (Figure 255).*
    This is an important step. If the center universal joint is not in a straight line with the two end joints, the shaft will not rotate at a constant speed. Poor alignment increases wear on the shaft bearings.

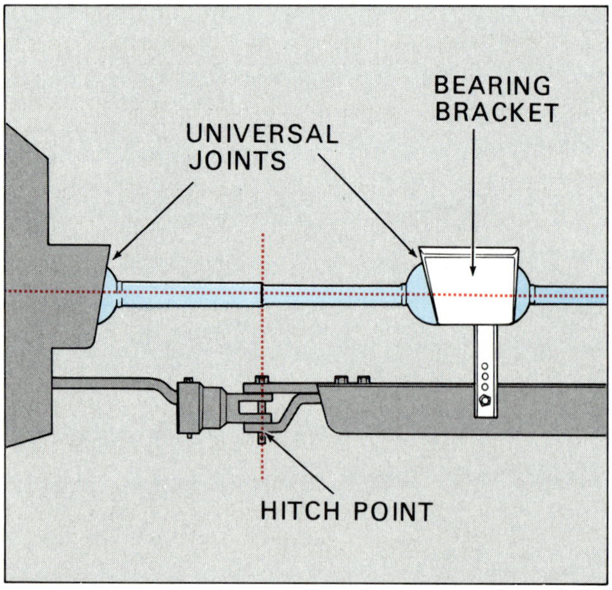

FIGURE 255. Adjust bearing bracket so PTO shaft is straight. Make sure hitch point is centered between universal joints.

13. *Attach safety shield over PTO shaft (Figure 256).*
    This step is particularly important because the usual PTO accident is the result of entanglement of the operator's clothing. The shield helps to avoid this possibility. In a study of 88 accidents involving the PTO shaft, one or all of the shields provided by the manufacturer were missing in 67 of the cases.
14. *Engage PTO.*
15. *Operate PTO at slow speed to check shaft connections.*
    Apply power easily. You can damage a machine by suddenly applying power if part of the mechanism is binding or is blocked by an obstruction.
16. *Check telescoping shaft(s) with tractor and implement in normal angle for turning corners.*
    If telescoping shaft shortens until it bottoms (will shorten no further), readjust bearing bracket to secure clearance. Incorrect adjustment can cause damage to the PTO shaft assembly.

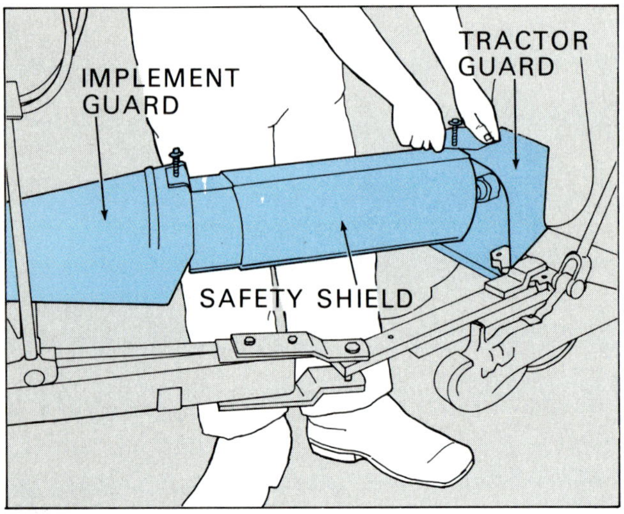

FIGURE 256. Be sure safety shield is fastened securely to the implement guard and to adapter guard on tractor.

6. CONNECTING THE REMOTE HYDRAULIC CYLINDER

The *remote cylinder* is one that can normally be removed from one implement and mounted on another (Figure 257). It is controlled separately by a second lever on the regular hydraulic control quadrant on the tractor. It is especially well fitted for controlling trailing implements or semi-mounted implements, but it can be used on tractor-mounted equipment in addition to the regular hydraulic hitch control.

The remote cylinder (sometimes called a "pull-behind cylinder," "portable cylinder," "hydraulic jack" or "hydraulic ram") is used mostly for such control jobs as lifting and lowering trailing types of implements and semi-mounted implements. Examples are plows, disk harrows and the cutter bars of combines and mowers. Another use is for reversing a 2-way tillage tool.

There are two types of remote cylinders—the **single-acting type** and the **double-acting type**.

A single acting cylinder exerts pressure in one direction only (Figure 258). However a double-acting cylinder can exert pressure at either end of the cylinder (Figure 259).

There are ASAE-SAE standard dimensions for both types; however, the standards have not been fully accepted by all manufacturers. Cylinders that meet these standards can be moved from one implement to another rather easily if the implements are designed for a standard cylinder.

FIGURE 257. A remote cylinder.

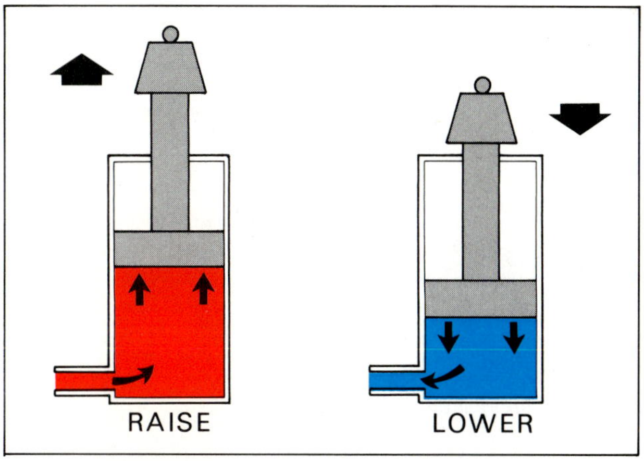

FIGURE 258. Working principle of a single-acting (closed-center) hydraulic system.

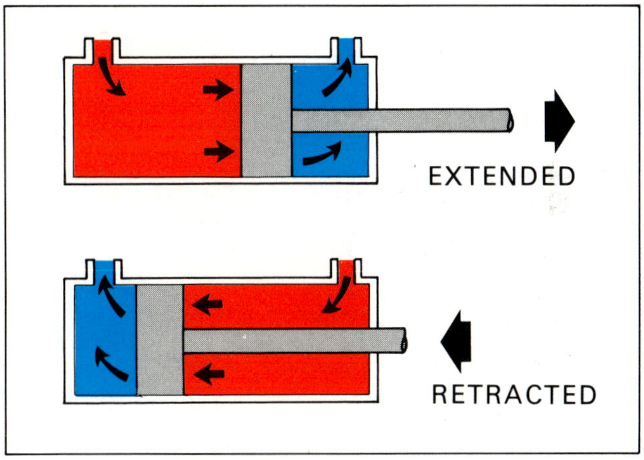

FIGURE 259. Working principle of a double-acting (open-center) hydraulic system.

The standards also provide for 3 categories of remote cylinders as follows:

Category	Max. Drawbar Power		Length of Stroke	
	kW	hp	cm	in
I	Up to 38	Up to 45	20	8
II	30 to 75	40 to 100	20	8
III	Over 60	Over 80	20 or 40	8 or 16

There are no standards for pressures to be supplied to the cylinders. Pressures vary with different makes of tractors from about 4823 kPa (700 psi) to more than 20,670 kPa (3,000 psi).

Not all remote cylinders are removable. Some are built on the implements they serve with the intention that they remain with the implements.

In general, the following procedures are used for connecting a **remote cylinder**:

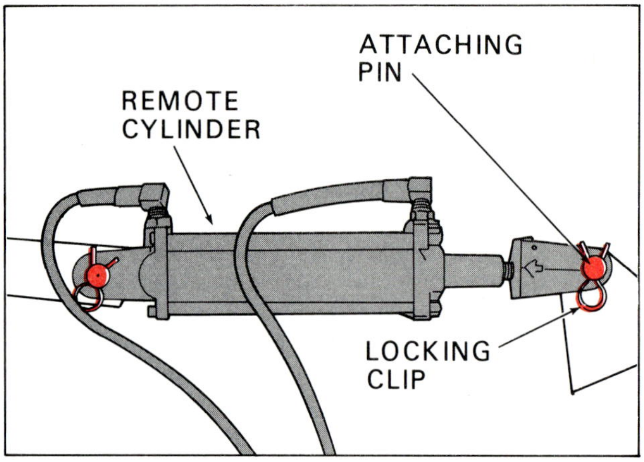

FIGURE 260. Remote cylinder is held at both ends with attaching pins. Attaching pins are held in place with spring locking clips.

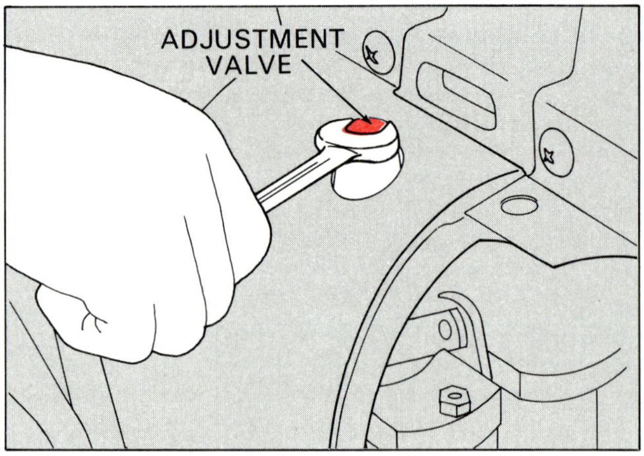

FIGURE 261. Some hydraulic systems can be adjusted to supply either single-acting or double-acting cylinders.

1. *Mount remote cylinder on implement (if not already mounted).*

   Mount in position so that hose length will not change when piston arm extends and retracts. The remote cylinder is attached at both ends with steel pins (Figure 260). Make certain that spring locking clips are inserted to hold steel pins in place.

   Some implements have a **locking device** to hold the implement in a raised position when the remote cylinder is removed. Before operating remote cylinder, remove locking pin or disengage latch if there is one on the implement.

2. *Examine cylinder to determine if it is single-acting or double-acting type.*

   If there is provision for only **one hose connection** on cylinder, it is **single acting**. If there are **two hose connections**, it is **double acting**.

3. *Determine if the tractor hydraulic system supplies one-way pressure or both.*

   Check your operator's manual. If one isn't available, examine the hydraulic outlets on the tractor. You may find one or more extending from the rear of the tractor or from either or both sides. If the outlets are grouped in pairs, the tractor probably supplies two-way pressure. Single outlets usually indicate one-way pressure.

   The hydraulic system on some tractors can be adjusted to supply either one-way or two-way action. This adjustment is accomplished by changing a valve on the control-valve body (Figure 261). You will need to check your operator's manual for instructions on this procedure.

4. *Remove dust plugs from couplings (Figure 262).*

   Rubber or metal plugs and caps protect the hydraulic connections from accumulating dirt and foreign material on the oily surfaces.

   The plugs (or caps) may be threaded and screwed

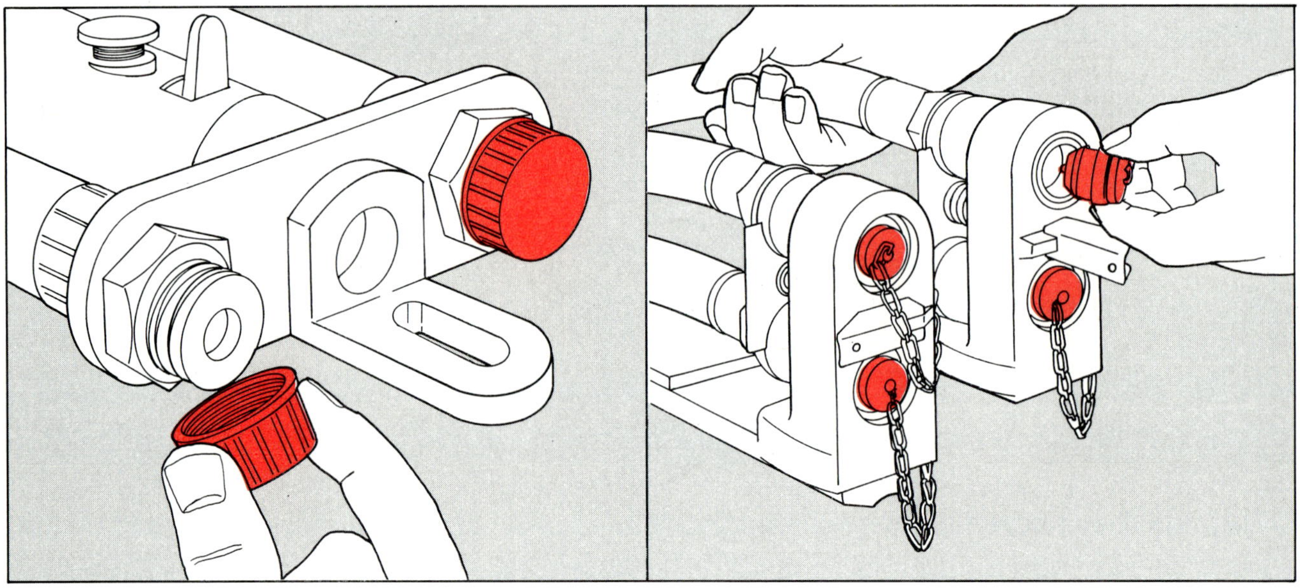

FIGURE 262. Types of hydraulic-coupling dust protectors.

FIGURE 263. Work control lever backward and forward to relieve pressure.

on the coupling or held by a gripping arrangement that requires pulling the hose towards you so that the coupling will release its grip on the plug, or if a rubber plug, it may be held by friction.

5. *Relieve pressure in tractor hose lines (Figure 263).*
   Be sure the engine is turned off. Work control lever backward and forward. Hold it in each position momentarily to release all pressure.
6. *Clean hydraulic connections and attach couplings.*
   Use a clean, lint-free cloth for cleaning connections, not an old sack or your fingers. Dirt enters the hydraulic system through dirty connections. It causes rapid wear of the pump, valves and cylinders, and results in an early repair bill.

Some hydraulic connections are joined by threads, but most hose lines connecting the remote cylinder to the tractor are of the "break-away" type (Figure 264). In case an implement becomes detached from the tractor, the pressure line will break connections without damage to the hose and without loss of oil.

If you have a yoke with two connections together, you may find the yoke and tractor couplings have matching letters or arrows to show proper connections. With a double connection of this type, provision is usually made for inserting a heavy screw driver or bar to force the connections into position. A latch holds them in position.

There is no uniform standard for couplings; consequently, the fittings of one manufacturer will not mate with those of another. If you change makes of tractors and the connections on it are different from those on your implement, the coupling on one or the other will have to be changed so that they are of the same make.

If your tractor supplies two-way hydraulic action and you have a single-acting cylinder, check your operator's manual to see which pressure connection to use. If a manual isn't availble, connect to either one. If you connected to the wrong one, in operating the control lever, the action of the machine will be opposite from that of normal operation. Changing the hose connection to the other outlet will give proper operation.

If you have a one-way hydraulic system and a double-acting cylinder, use the hose that connects to the end opposite the piston rod—it has more piston area and will lift more weight than if the connection is made to the one supplying the piston-rod end.

On some tractors, the hose that isn't used for cylinder operation can be connected to feed back into the hydraulic system where it will drain and be kept free of dirt. If your tractor doesn't have this provision, drain the oil from the side that won't be used and insert a **breather filter** (Figure

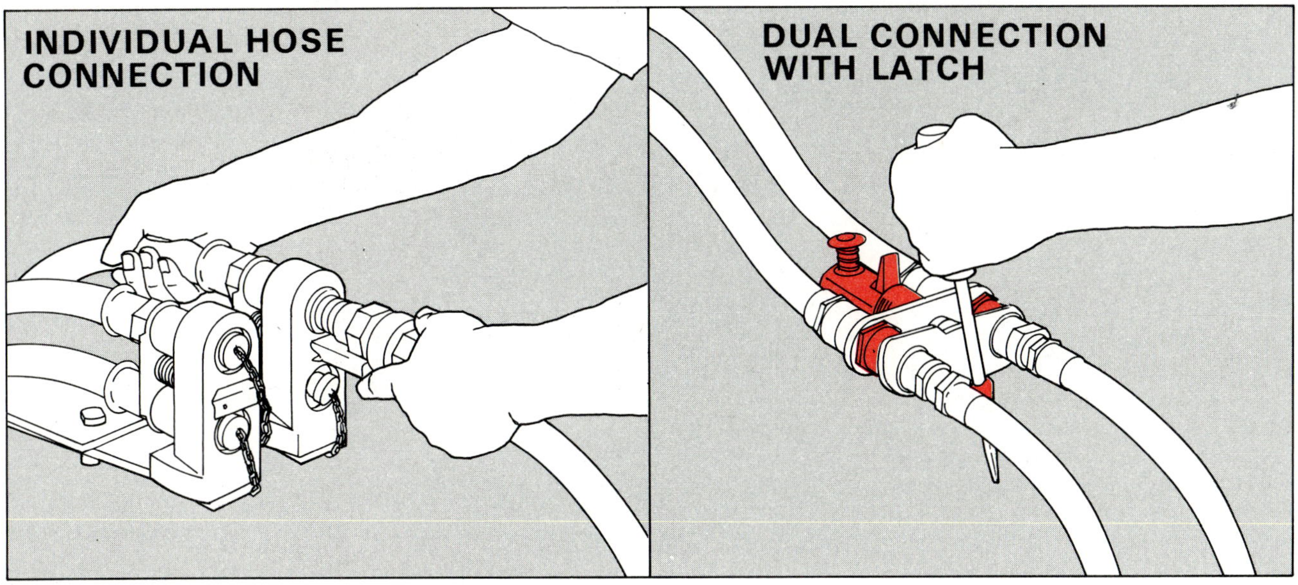

FIGURE 264. Break-away types of hose couplings.

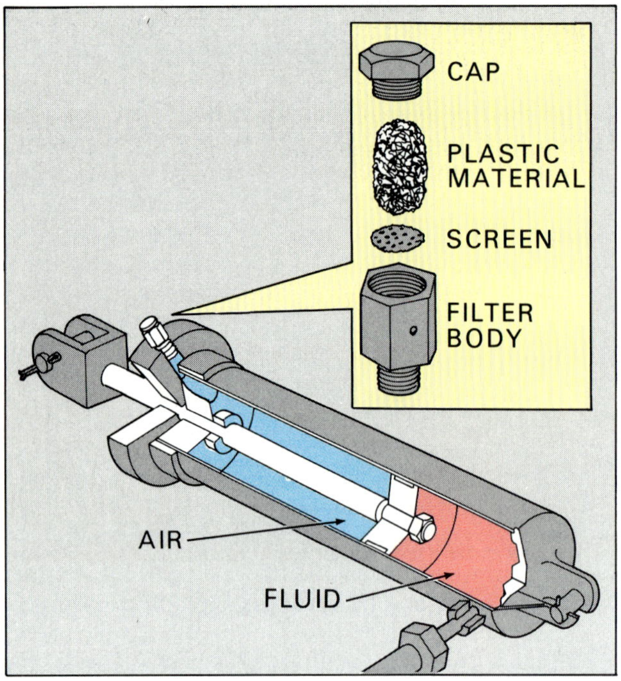

FIGURE 265. Breather filter for unused port of double-acting cylinder when used with single-acting hydraulic system. Filter allows free air passage but prevents dirt from entering.

265). The breather filter allows air to enter and discharge as the piston retracts and extends but keeps dirt from entering.

Don't leave the oil port open and exposed to dirt. Enough dirt will soon enter to ruin the cylinder. Some tractor operators cap the opening, but this causes the piston to compress the trapped air when it extends so you can't get a full stroke. If capped while the piston is extended, a vacuum develops which limits the length of the stroke when the piston retracts.

7. *Adjust hydraulic-hose support so that coupling ends will not touch ground when disconnected.* If the implement should become disconnected from the tractor, the break-away couplings work best if the hose lines pull straight back on the coupling. A downward or upward pull may keep the coupling from separating. Provide for this action in the adjustment of the hose support (Figure 266).

8. *Operate hydraulic cylinder and check for proper operation.* As indicated in step 5, **if the implement moves in the wrong direction** when operating the control lever, exchange the hose connections on a two-way hydraulic system.

   If the **cylinder doesn't move smoothly**, you may have air trapped in the oil line. In that case, remove the cylinder and set it on end with the hose connections on the upper end. Move the control lever back and forth seven or eight times, extending and retracting the piston until all air is removed.

FIGURE 266. Hydraulic-hose support should be adjusted so that hose lines pull straight on couplings if implement becomes disconnected from tractor. This provides for easy break-away of couplings.

9. *Check hydraulic-system oil level (Figure 267).* Operating the cylinder a few times before checking is important to make certain that pockets of air are removed.

   If your tractor has a **single-acting hydraulic system**, extend the piston before checking the oil level to be sure you have ample reserve.

   Too little oil may cause the hydraulic system to overheat. Overfilling will also cause oil heating and oil foaming. Check your operator's manual for the **proper oil** to use. Recommendations vary with different tractors.

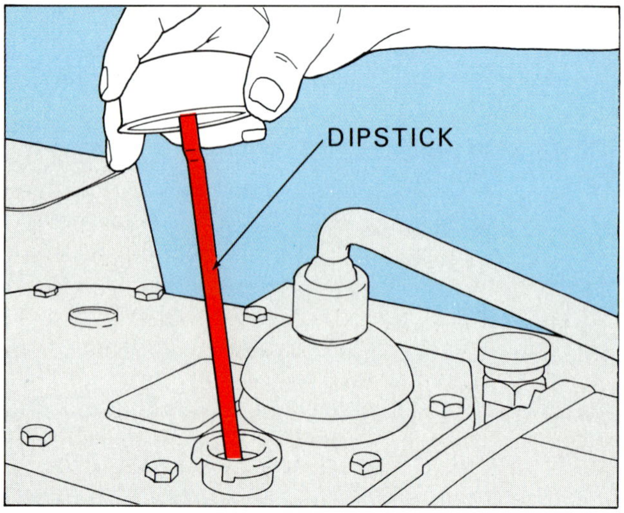

FIGURE 267. Hydraulic system oil-level dip stick.

7. HITCHING TO BELT-DRIVEN EQUIPMENT

Power for belt work was once one of the major jobs for farm tractors. At that time, most tractors had the pulley mounted on the side of the tractor. However, in the last few years, use of the belt pulley as a power source has decreased to the point that pulley attachments are now available only as special units which attach to power take-off drives (Figure 268). The pulley attachment is used only when belt power is needed and removed when the tractor is used for other purposes. Seldom is a side-mounted belt pulley provided on late-model tractors. Because of this, hitching to belt-driven equipment is not discussed in this text. Check your operator's manual or consult with your local tractor dealer if you wish to use belt-driven equipment.

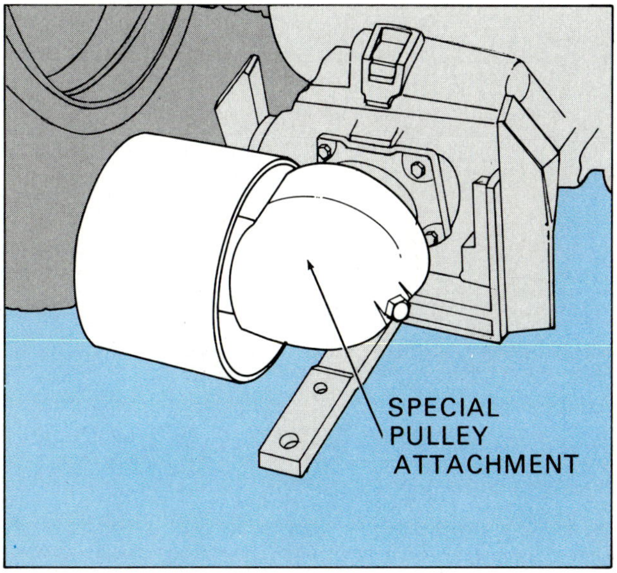

FIGURE 268. Pulley mountings.

B. Unhitching Equipment

Unhitching implements from the tractor is mostly a matter of reversing the procedures for connecting implements. Since detailed procedures were given for connecting the hitch, PTO and hydraulic equipment, only general procedures and precautions are given here for disconnecting.

Upon completion of this section you will be able to unhitch equipment properly and safely.

If you have **hydraulic and PTO connections** in addition to the hitch, use the following sequences:

1. *Place implement where it will remain stable after hitch is disconnected.*
2. *Disconnect hydraulic connections first (Figure 269).*

   Be sure to **cap ends of hose couplings** to keep them from collecting dirt. If you drop the connections on the ground, be sure they are wiped clean before capping, in order to avoid the possibility of dirt's entering the hydraulic circuits when the lines are reconnected. A small amount of grit can do a large amount of damage to the hydraulic pump.

   The remote cylinder can be left on the implement if desired. If removed on an implement such as a plow, **be sure the safety pin is in place** before disconnecting the hydraulic hose. Both breakaway and threaded-type hydraulic connections between the implement and tractor are provided with check valves to prevent loss of oil. When the disconnection is complete, only a few drops of oil escape.
3. *Disconnect PTO shaft at tractor.*

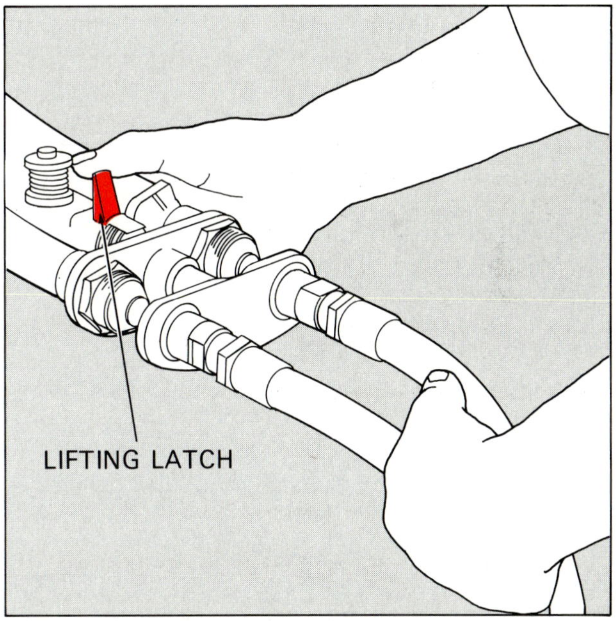

FIGURE 269. Lifting latch releases yoke holding hydraulic connections. Some types are released by pulling the tractor hose forward.

   Be sure the PTO shaft is covered. Use a master guard, tube or cap. Don't neglect this point.
4. *Disconnect hitch.*

   If the implement is not equipped with jacks to maintain it in position for rehitching, rest the implement on wood blocks so that it can be reconnected easily (Figure 270).

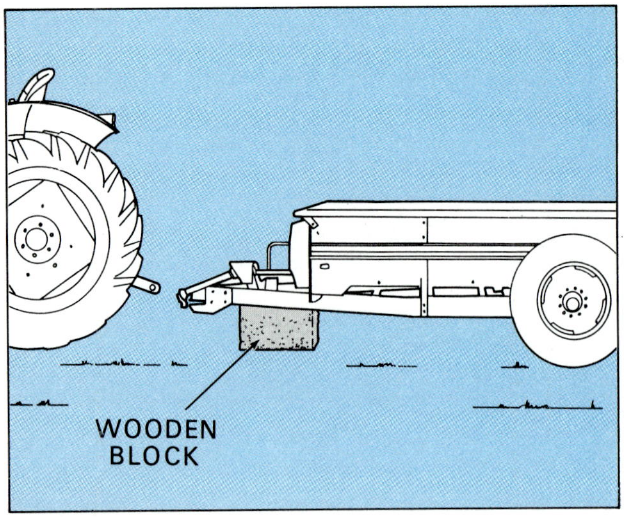

FIGURE 270. Rest implement on wood block for easy re-connection.

VI. Operating and Transporting Tractors

At this point, you should have already mastered the skills of servicing and driving the tractor. Now you will learn how to operate it with implements and under field conditions.

From your study of this section, you will be able to **operate the tractor safely with implements attached and under load**. These processes are discussed under the following headings:

A. Operating Tractors Under Load.
B. Transporting Tractors and Equipment.

A. Operating Tractors Under Load

Upon completion of this section you will be able to operate tractors under load.

Procedures are given under the following headings:

1. Preparing for Operation.
2. Checking the Hydraulic System Operation.
3. Matching Gear Selection and Engine Speed with Load.
4. Handling Overloads Without Stopping Tractor.
5. Checking and Correcting Tire Slippage.

1. PREPARING FOR OPERATION

Before operating your tractor and implements, think about what you would do in case of an emergency. Once an accident happens, you seldom have time to study the situation. Check tractor and implements for proper operation.

1. *Make safety check (Figure 271).*
2. *Adjust row-width spacing, if necessary.*
3. *Make a few turns around the yard before going to the job.*
4. *Check steering, brakes and hydraulic system.*
5. *Check engine operation—instruments such as ammeter and heat indicator.*
6. *Check operation of implement.*

Safety Checklist for Operating Tractors and Equipment

- ☐ *Read and understand operators manual.*
- ☐ *Wear proper clothing.*
- ☐ *Be in good physical condition.*
- ☐ *Know your hand signals.*
- ☐ *Check visibility.*
- ☐ *Check fire extinguishers.*
- ☐ *Check first aid kit.*
- ☐ *All guards and shields in place.*
- ☐ *SMV emblem properly displayed.*
- ☐ *Area clear of obstacles, people and animals.*
- ☐ *Seat belts fastened (with ROPS).*
- ☐ *No riders.*
- ☐ *Watch for pinching, cutting and crushing hazards.*
- ☐ *Equipment securely and properly attached.*
- ☐ *Equipment calibrated.*
- ☐ *Row widths adjusted.*
- ☐ *Watch for obstructions—overhead, ground and side.*

FIGURE 271. Use this safety check list to check tractor and implement for proper operation before going on the job.

2. CHECKING THE HYDRAULIC SYSTEM OPERATION

The hydrauiic system of most tractors operates when the engine is running. Hydraulic power is needed continuously for implement control and to supply power for brakes, steering, hydraulic motors, etc.

There is a provision in most hydraulic systems for relieving the pump automatically when hydraulic demands are low.

More complete details on how hydraulic systems work are given in the AAVIM publication, "Hydraulics for Mobile Equipment."

If you plan to use the *hydraulic system* of your tractor, proceed as follows:

1. *Move hydraulic-control lever through "raise" and "lower" positions to make certain system is operating satisfactorily.*
   As you operate the hydraulic controls of different tractors, you will find there are two different control actions—one is the **limit control**, the other a **proportioning control** (Figure 272).
   With the **limit control**, movement of the control lever in either direction from neutral will cause the implement to continue rising (or lowering) until the lever is returned to neutral. The implement remains fixed in that position until the control lever is moved again.
   With the **proportioning control**, movement of the control lever in either direction will cause the implement to be raised or lowered in proportion to the distance you move the lever on the quadrant.

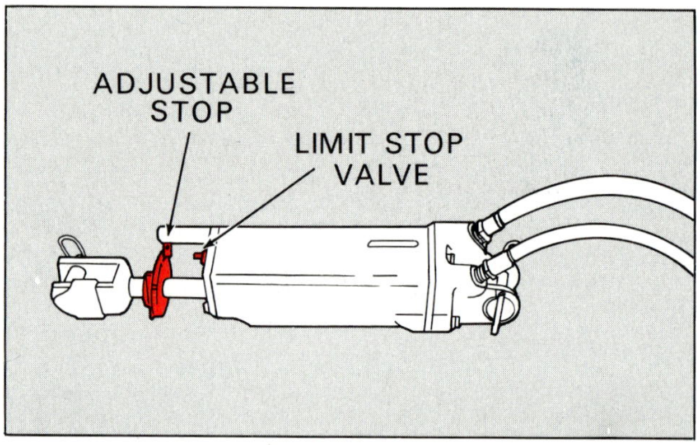

FIGURE 273. Adjustable stops used on cylinders to control depth of soil-engaging implements and height of lift of other implements.

2. *Set supplemental hydraulic controls as necessary.*
   If your tractor is connected to a **trailing implement which is equipped with a remote cylinder**, set stop on piston rod for approximate stroke desired (Figure 273). For plows, this limits the plowing depth. For grain drills, combines and implements where the cylinder is used for lifting, it determines the height of lift. You will probably have to readjust the stop after you try your equipment on the job.
   If your tractor is connected to a **mounted implement**, it may have a similar cylinder and adjustable-stop arrangements as used on the remote cylinder (Figure 272).
   For direct manual control of the hydraulic unit, there may be a stop located on the quadrant. When the soil-engaging equipment has reached the proper depth, you can set the stop so that the

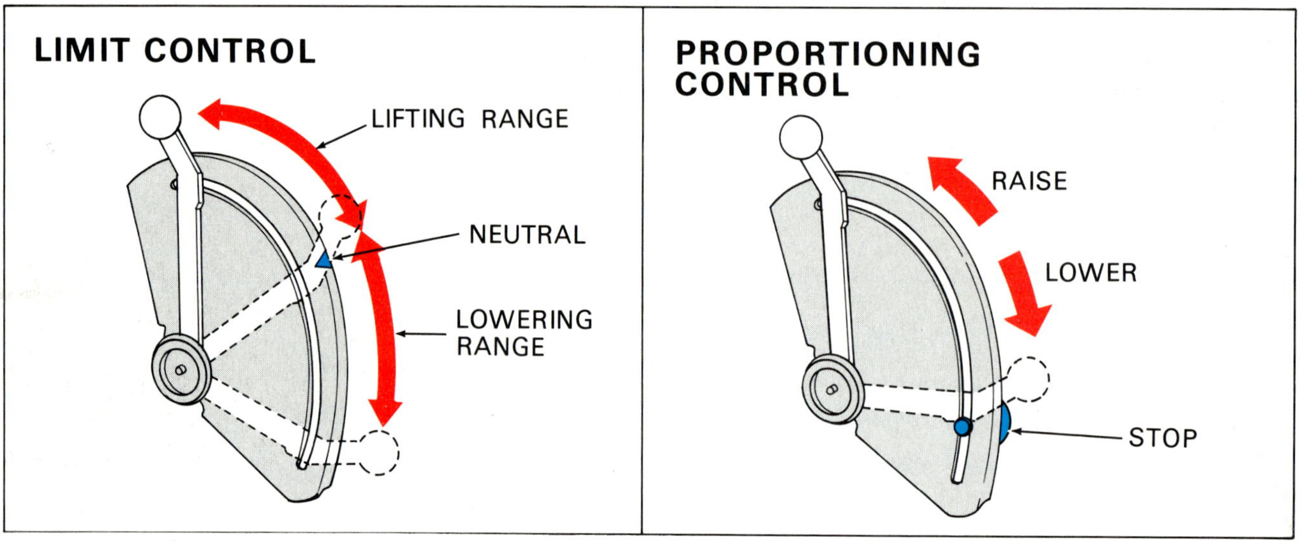

FIGURE 272. Principles of how limit controls and proportioning controls work.

FIGURE 274. Implements with gage wheels control their own depth independently of the hydraulic control. (This one also serves as a drive wheel and press wheel.)

FIGURE 275. Draft controls are used with mounted, soil engaging implements such as plows and cultivators.

lever can be returned to the same position to assure the same depth when the implement is again raised and lowered. This feature is sometimes called "position control," "implement position control" or "uniform depth control."

If you have a **mounted implement with its own gage wheel** (Figure 274), set your hydraulic-control lever in **"free-floating" position**. The gage wheels control implement depth and need to be free to raise and lower in accordance with the changes in surface conditions.

The "free-floating" position of the hydraulic control varies with different tractors. Check your operator's manual for proper positioning of the control lever.

Some tractors are equipped with **automatic draft control**. This is a weight-transfer device built into the tractor hydraulic system. It is used with mounted implements of the soil-engaging type such as plows and cultivators (Figure 275).

When you wish to use draft control, there is a separate lever to engage. It may be mounted on the gear case or on the quadrant assembly along with the hydraulic-control lever (Figure 276). Draft controls vary somewhat in the way they are adjusted so be sure to use the operator's manual for your tractor for exact instructions (See Hydraulics, Volume I, AAVIM, for details).

(Draft control is known by such trade names as "constant draft control," "load control," "traction booster" and "tractor amplification.")

3. *Check rate of lowering and raising of implement or hitch.*

A flow control valve, commonly called a "response control" (Figure 277), governs the speed with which an implement is lowered for ground engagement. For plowing or cultivating, a fast

FIGURE 276. Means of engaging automatic draft controls.

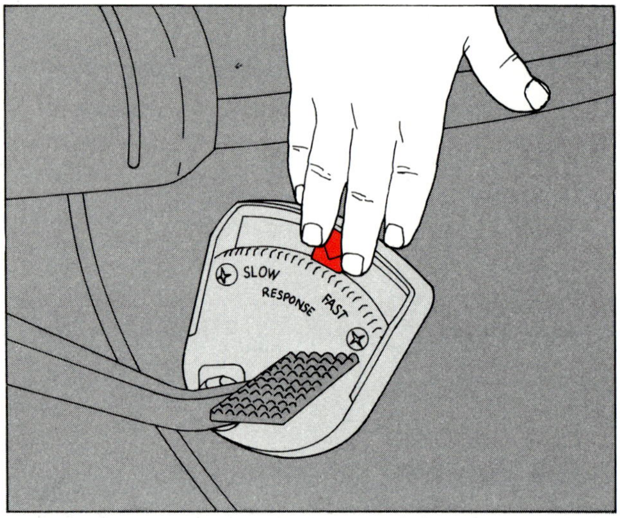

FIGURE 277. Response control which provides for faster or slower lowering and raising of implements or for adjustment response to soil-engaging tools.

drop is desirable while a slow gentle drop is recommended for a planter. Similarly, it will make adjustments for the speed of lift as well as for corrections to draft response for soil-engaging implements. Some tillage tools penetrate quickly and then withdraw momentarily to cause a "bobbing" action. Moving the control to a slow response corrects this action.

4. *Check implement hitch.*

   Some implements provide two positions for the tongue. One position is to center the implement back of the tractor for transporting.

   The other position is for field use and provides for positioning the implement to one side of the tractor hitch. These are common on field balers, combines and corn harvesters.

3. MATCHING GEAR SELECTION AND ENGINE SPEED WITH LOAD

Some operator's manuals give suggested speeds for various jobs. For example, if you have four speeds forward on your tractor, your selection can be approximately as follows:

- **1st — For extra-heavy or slow-speed work.**
- **2nd — For plowing.**
- **3rd — For cultivating plants that are so large that they don't cover easily.**
- **4th — For light work.**

If you have five speeds forward, the first and second gears are used for heavy work.

These are good general recommendations, but the conditions under which you operate a tractor vary widely. Often you need to know more about what is involved in selecting a gear that will get your field job done effectively and still use an engine speed that will give satisfactory engine efficiency.

You get greatest efficiency from a **carburetor engine** on a **heavy load** when the speed-control lever is completely open. On a job that requires full power occasionally, select a gear that will allow you to operate with the speed-control lever completely open in order to keep the reserve power of the engine available for times when you need it. (Some tractors have special means for handling overloads built into the tractor. These are discussed under the next heading.)

If you are operating a carburetor engine at light load, you will get the greatest efficiency by using the highest gear at which the tractor will still have the necessary pull. The Illinois study showed that operating a tractor in this manner (as compared to operating it with the speed-control lever fully open and at a lower gear ratio) showed savings as high as 21.9 percent in the amount of fuel used.

None of the tests reported used a throttle-lever setting less than ½ open. It is rather generally accepted that an engine should not be operated on continuous pulls with the speed-control lever much less than ½ open.

If you are using a **diesel engine** on a **heavy load**, it, too, operates with the most efficiency when fully loaded with the speed-control lever completely open.

A diesel tractor is noted for its ability to "lug" under these conditions; but if it is about to stall under heavy load, the engine may start running backwards. When it does, air is drawn in through the exhaust valves and forced out through the air intake and air cleaner. Besides getting an oil bath from oil blown out of the air cleaner stack, it is extremely dangerous. Diesel engines will develop almost full speed when the engine reverses if the speed-control lever is completely open.

For **light loads**, follow the same procedures as with a carburetor engine. A diesel engine has the advantage of higher efficiency at part load than a carburetor engine and has better lugging ability at slow engine speeds. But it is important to keep engine speed high enough to avoid excessive exhaust smoke.

Don't overload either type of engine or load it so heavily there is no reserve for momentary overloads. One manufacturer suggests this test for checking reserve power of a carburetor engine: "With the tractor in motion, set the throttle half open. Then quickly pull the throttle fully open. If the tractor speeds up rapidly, the engine is not overloaded."

If you are using the **PTO**, set the speed-control lever to supply the proper PTO speed, then adjust ground speed with your gear selection.

If your tractor is equipped with a **tachometer** (an instrument which indicates engine speed in revolutions

FIGURE 278. Tachometer (for indicating engine speed) is often included with an hour meter (for showing number of hours engine operates at a standard speed). Combination instruments of this type are known by such names as "proofmeter," "speed-hour meter," "operation meter" and "tachourmeter."

per minute), the proper engine speed for the PTO is marked on it (Figure 278). If it has no tachometer, check your operator's manual—some tractors provide proper PTO speed when the speed-control lever is fully open. Another method is to check a slow-moving part, such as the straw rack on a combine, to determine when the tractor engine is operating at proper speed. Instructions for the machine indicate the proper straw rack speed. Otherwise, you will need to check speed at the end of the tractor PTO shaft with a speed counter.

4. HANDLING OVERLOADS WITHOUT STOPPING TRACTOR

As pointed out in the previous discussion, you get the greatest efficiency from your tractor if you can operate it near its full-load capacity. The problem is that any extra demand for power creates an overload which, if continued, can be very hard on your tractor. Also, it may cause you to kill your engine at a time when it is difficult to start the load again from a dead stop.

To meet these conditions, manufacturers have developed a number of ways to "down shift" while your tractor is still moving. This action supplies the extra drawbar power needed to meet most temporary overload conditions.

Tractors with power-assist shift transmissions—discussed under "Identifying the Type of Tractor Transmission"—already have provision for down shifting on-the-go. Consequently, this discussion does not apply to them.

FIGURE 279. A planetary drive lever.

Tractors with **standard gear transmissions**—those which must normally be stopped to change gears—are the ones that use the **overload units** to advantage. The units enable the operator to slow ground travel without stopping and at the same time increase drawbar pull from about 40 percent to as much as 100 percent. These units are also good for starting extra heavy loads. They are of three different types:

- **A planetary gear unit located between the engine clutch and transmission.**
- **A double clutch with reducing gear arrangement located between engine clutch and transmission.**
- **A torque converter located between the engine and engine clutch.**

The **planetary units** go by such commercial names as "ampli-torc" and "torque amplifier." When you are operating a tractor equipped with this unit and encounter an overload, you simply pull back on the planetary-drive lever without stopping the tractor or disengaging the clutch (Figure 279).

When you engage the planetary unit, you immediately get a slower forward speed and an increase in drawbar pull. When the overload condition is past, the planetary-drive lever is pushed forward so the tractor again travels at its original speed for a more economical operation.

The planetary unit can be used in any of the forward or reverse speeds.

The **double-clutch unit** is known by such commercial names as "power director," "hydra-power" and "multipower." It works in much the same way as the planetary unit, from an operator's standpoint. The principal difference is that some have a neutral position. Neutral is used to stop forward movement of the tractor.

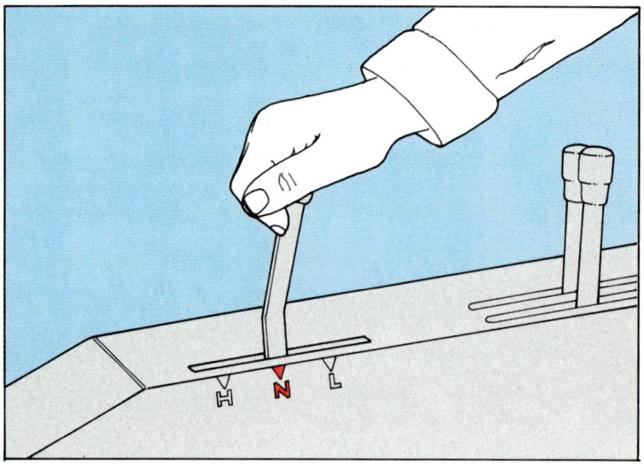

FIGURE 280. Double clutch in neutral.

As shown in Figure 280, when the control lever is in "neutral," both clutches are disengaged. The tractor has no forward movement.

When the control lever is pushed forward (Figure 281), the "high-range" clutch engages. Power flows from the engine directly to the transmission shaft. The rest of the mechanism rotates, but transmits no power.

When an overload develops, the control lever is pulled back (Figure 282). This automatically disengages the high-range clutch and engages the "low-range" clutch. Power from the engine feeds through the speed-reducing gears to provide slower speed to the transmission shaft. Ground speed is slowed, but more pull is provided at the drawbar to meet the overload condition.

There are several designs of double clutches for handling overloads. For example, some are controlled by mechanical linkage while others are controlled through the hydraulic system. On some, "neutral" is not used. Some units use an "overdrive" clutch for high range.

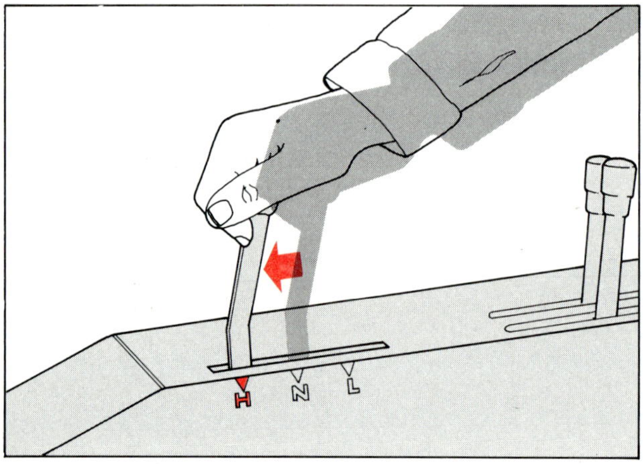

FIGURE 281. Double clutch in high range.

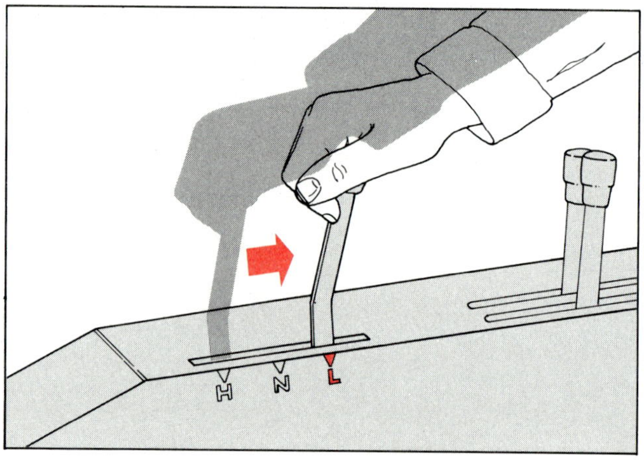

FIGURE 282. Double clutch in low range.

The **torque converter** is similar to those used in the automatic transmissions of automobiles. They are known on tractors by such commercial names as "lugmatic" and "powermatic" (Figure 283).

There is no lever to move when you use a torque converter for overloads. As an overload develops, the turbine speed (and tractor ground speed) slows in approximate proportion to the overload. At the same time, drawbar pull increases. When the overload is past, tractor ground speed increases to normal and drawbar pull becomes less.

Some torque-converter tractors have a lever which locks the converter to the flywheel. This action cuts out the converter and provides a direct drive for use with light loads or for high speeds. But when you press the clutch to stop the tractor, the torque converter is automatically unlocked. When you again engage the clutch to start movement, the torque converter is engaged.

FIGURE 283. Torque converters are often recognized on modern tractors by their commercial names.

FIGURE 284. Tread wear of this type is caused by excessive wheel slippage.

5. CHECKING AND CORRECTING TIRE SLIPPAGE

Some tire slippage is to be expected with rubber tires, but it is generally agreed that when slippage becomes greater than 15 percent during normal tractor operations you should either add weight to the rear wheels to provide more traction or reduce the tractor load. Because of rapid tire wear, this percentage figure is too high when operating on such surfaces as concrete. It is difficult to see slippage until it reaches about 15 percent.

Slippage is important to you for three reasons: (1) it causes rapid tire wear (Figure 284), (2) wastes fuel and (3) wastes valuable time.

When there is heavy drawbar pull and traction conditions are poor, one drive wheel may start to spin. Slippage often occurs to the land wheel while plowing because most of the weight is on the furrow wheel due to the tilt of the tractor.

You can reduce slippage by one of the following methods:

- **If plowing, add more weight to the land wheel.**
- **Engage the differential lock.**
- **Apply braking action on the spinning wheel to slow it down (temporary action only).**

Additional weight on the land wheel of the tractor is recommended only when plowing. The additional weight should be removed when not needed.

A **differential lock** is available as either standard or optional equipment on many tractor models. If your tractor has a differential lock and one wheel starts to slip, engage the hand or foot lever that controls the lock. The locking device is in the differential. It locks both rear axles so they drive as a solid shaft. Power is applied evenly to both wheels, regardless of slipping or turning conditions. The wheel with the best traction will then increase its pulling effort without the opposite wheel spinning.

In most units, the clutching device will release automatically when wheel pull equalizes. If it doesn't release, you may have to engage the engine clutch or the brake depending on the type unit you have. Check your operator's manual to learn how your unit works.

These units may be either mechanically or hydraulically operated.

Note: When turning, make certain the differential lock is released. When a tractor turns, the outside wheel must rotate faster than the inside wheel because it has a greater distance to travel. If the differential lock is still engaged, both wheels are forced to rotate at the same speed and on turns one must slip. This action causes excessive strain on the differential and rapid tire wear.

If you apply **braking action** on the spinning wheel, it will slow down. Greater pulling effort is transferred to the wheel that has good traction. Under most conditions, tractor travel will continue. However, wheel braking causes brake wear, wastes engine power, strains the drive mechanism and may even stall the engine. Use this method only for emergencies or for short periods of time.

Some tractor operators reduce tire slippage in sand, mud or snow by lowering the pressure in both rear tires. However, don't be misled by what appears to be an advantage in using low tire pressures. Although studies show that traction is increased, you are taking two serious risks: (1) the possibility of slipping a tire on its rim and ruining a tube and (2) the risk of cracking and ruining the sidewall of a tire from buckling. Only in extreme emergencies where increased traction is needed temporarily can you justify the risk of losing a set of rear tires to get more traction.

To check tire slippage, proceed as follows:

1. *Disconnect tractor from load.*
2. *Put chalk mark on tire directly below axle (Figure 285).*

FIGURE 285. To check tire slippage, mark tire on drive wheel with chalk.

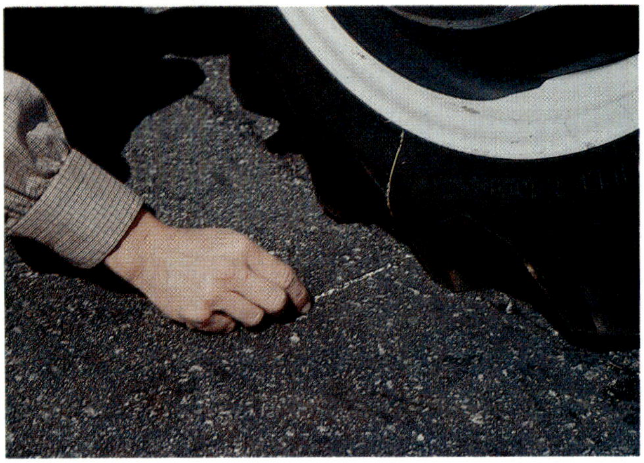

FIGURE 286. Mark the ground opposite tire mark.

3. *Mark ground or set stake in ground next to chalk mark (Figure 286).*
4. *Drive tractor straight ahead until tire makes exactly 10 revolutions (Figure 287).*
   Be sure chalk mark is directly under axle when you stop. This is point "A."
5. *Measure distance from starting point to point "A."*
   This is to determine the actual distance the tractor travels at ten revolutions of the drive wheel without slippage.
6. *Connect tractor to load.*
7. *Repeat steps 2 through 4.*
   Start tractor at original starting point. The location where the chalk mark stops after 10 revolutions of the tire is point "B" (Figure 288).
8. *Measure distance from starting point to point "B."*
   For field work, mark off distances determined in step 5 and follow the same procedure.
   Figure percent of slip as follows:

$$\frac{\text{(distance to) A} - \text{(distance to) B}}{\text{(distance to) A}} \times 100 = \begin{array}{c}\text{percent}\\ \text{of slip}\end{array}$$

Example using distances from Figures 287 and 288:

SI:

$$\frac{36.5 - 29.8}{36.5} \times 100 = \begin{array}{c}\text{18.3 percent}\\ \text{slip}\end{array}$$

U.S.:

$$\frac{120 - 98}{120} \times 100 = \begin{array}{c}\text{18.3 percent}\\ \text{slip}\end{array}$$

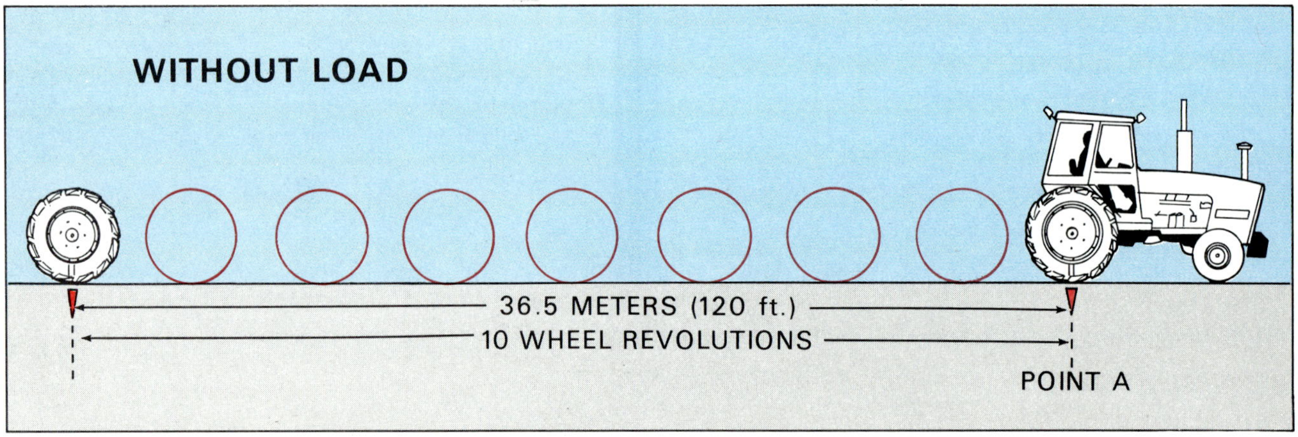

FIGURE 287. Without load, drive tractor 10 revolutions of drive wheel and mark on ground opposite chalk mark (point A).

FIGURE 288. With load, drive tractor 10 revolutions of drive wheel and mark on ground opposite chalk mark (point B).

B. Transporting Tractors and Equipment

Whenever possible, mount the tractor and equipment on a truck or trailer for transporting. Proceed as follows:

1. *Always back the tractor onto the trailer (Figure 289).*
   This prevents tractor from overturning.

FIGURE 289. Back a tractor on a trailer when loading it for transport.

2. *Secure the tractor and equipment with chains (Figure 290).*
   Use load binders to lock the tractor and equipment down.
3. *Connect trailer lights (Figure 291).*

FIGURE 290. Chain tractor and equipment securely to trailer.

FIGURE 291. Connect all lights before towing trailer.

NOTES

Index

Q

R

S

T

U

V

W

X

Y

Z

Acknowledgments

INDIVIDUALS

Bear, Forrest
Minnesota Curriculum Services Center, White Bear Lake, MN

Beggs, H. Cecil
Agricultural Engineering, University of Georgia

Blanton, Lloyd
Agricultural Education, Clemson University, SC

Bruwelheide, Kenneth L.
Agricultural & Industrial Education, Montana State University

Coleman, Benjamin
Student, University of Georgia

Espenschied, R. F.
Agricultural Engineering, University of Illinois

Farmer, Gary
Georgia State Department of Education, Athens, GA

Fleming, Frank
Fleming's Farm, Athens, GA

Forbes, Allan
Leadership Institute, Holland College, PEI, Canada

Grant, Lee
Agricultural Engineering, University of Maryland

Halcomb, Alvin
Alabama State Department of Education, Auburn, AL

Hansen, Herb
Agricultural Engineering, Oregon State University

Harrell, Patrick
Agricultural Engineering, University of Georgia

Hayles, Jasper
Agricultural Education, Arkansas State University

Herd, Amon
Instructional Materials Lab, University of Missouri

Hoerner, Tom
Agricultural Engineering, Iowa State University

Howell, Dave
Agricultural Education, Penn State University

Jacobs, Clinton O.
Agricultural Education, University of Arizona

Kelley, Van
Agricultural Engineering, South Dakota State University

Kimmons, Marion
Agricultural Engineering, West Virginia University

Mason, J. Philip
Agricultural Engineering, Virginia Tech

Meyer, Charles
Student, University of Georgia

Moncrief, James. B.
United States Golf Association Green Section, Athens, GA

Page, Foy
Vocational Instructional Services, Texas A&M University

Perkins, Dana L.
Area Teacher for Vocational Agriculture, Athens, GA

Pierce, Greg
Oklahoma State Department of Education, Stillwater, OK

Pruitt, Pat
Agricultural Education, Utah State University

Rice, Charles E.
Agricultural Engineering, University of Georgia

Richardson, Joe
Agricultural Institute, New Mexico State University

Ridenour, Harlan
Ohio Agricultural Education Curriculum Materials Service, The Ohio State University

Silletto, Tom
Agricultural Engineering, University of Nebraska

Simpson, Gini
Student, University of Georgia

Simpson, Broaddus
Miller's Farm, Iron City, GA

Slyter, Damon
Kansas State Department of Education, Topeka, KS

Strawser, James D.
Cooperative Extension Service, University of Georgia

Sullivan, Roger
S & K Bike Shop, Inc., Athens, GA

Tart, C. V.
N. C. Department of Public Instruction, Raleigh, NC

Taylor, Douglas
Alberta Department of Agriculture, Edmonton, Alberta, Canada

INDUSTRY

A & I Equipment Company
Atlanta Highway, Athens, GA 30606

Allis Chalmers Corporation
Box 512, Milwaukee, WI 53201

Beall Peanut Company
Donalsonville, GA 31745

Bullock Ford Tractor, Inc.
235 Oak Street, Athens, GA 30601

J. I. Case Company
700 State Street, Racine, WI 53404

John Deere Waterloo Works
Technical Services, P. O. Box 270, Waterloo, IA 50704

Deutz Corporation
7585 Ponce de Leon Circle, Atlanta, GA 30340

Ford Tractor Operations
Ford Motor Company, 2500 East Maple Road, Troy, MI 48084

Hammond Tractor Company
Route 3, Monroe, GA 30655

International Harvester Company
AE Product Support Center, Technical Publications, P. O. Box 511, Hinsdale, IL 60521

Logan Equipment Company
1000 Winterville Road, Athens, GA 30605

Long Manufacturing
P. O. Box 1139, Tarboro, NC 27886

Massey-Ferguson, Inc.
1901 Bell Avenue, Des Moines, IA 50315

New Idea
AVCO, Coldwater, OH 45828

White Farm Equipment Company
Technical Publications, P. O. Box 701, Libertyville, IL 60048

NOTES

NOTES